GEMS MINERALS CRYSTALS *and* ORES

GEMS MINERALS CRYSTALS *and* ORES

The Collector's Encyclopedia

BY RICHARD M. PEARL

GOLDEN PRESS · NEW YORK

ACKNOWLEDGMENTS

In addition to the contributors thanked in the Preface, the author wishes to express his appreciation to the following for photographs supplied: Ward's Natural Science Establishment, for the snowflake obsidian, sandstone, trachyte porphyry, biotite gneiss, meteorite, shoulder-strap knapsack, four-drawer cabinet, glass-walled showcase, double-tiered display cabinets; Frantom Lapidary, for the large slab saw; Highland Park Manufacturing Co., for the lapidary's polishing and buffing machine, table-model lapidary outfit, lapidary's trim saw; Felker Manufacturing Co., for the diamond-set blades; Rocks Lapidary Equipment, for the cabochon-making unit; Vi-Bro-Lap Co., for the vibrating grinder and polisher; Smithsonian Institution, for the flat surface polisher; Geode Industries, Inc., for the rotating and vibrating tumblers; Arrow Profile Co., for the portable faceting unit; Estwing Manufacturing Co., for the rock-hunting tools; Star Engineering Co., for the field rock saw; DeBeers Consolidated Mines, Inc., for the three stages in diamond cutting; Sinclair Oil Corp., for the geologic time scale.

1977 Edition

 Printed in the U.S.A. by Western Printing and Lithographic Company. Published by Golden Press, Inc., New York, N.Y.

Dedicated to

Dr. Ben Hur Wilson

Geologist, Teacher, Author

Editor of *Earth Science*

ABOUT THE AUTHOR

Richard M. Pearl, a graduate of the University of Colorado and Harvard University, is Professor of Geology at Colorado College. The author of twenty-one books about the mineral world (translations have appeared in French, Japanese, Russian, and Persian), he is a contributor to four encyclopedias and has writtten more than 300 articles for scientific, trade, and popular periodicals. International recognition has come to him from gem and mineral organizations in thirteen countries. Professor Pearl is a well-known lecturer to both laymen and science teachers and has been active as founder or co-founder and past-president of the American Federation of Mineralogical Societies, the Rocky Mountain Federation of Mineralogical Societies, and the Colorado Mineral Society. He has collected and prospected in nearly every state and province in the United States and Canada. A Fellow of the American Association for the Advancement of Science, the Meteoritical Society, and the Gemmological Association of Great Britain, he is a past-president of the Southwest Section of the National Association of Geology Teachers.

CONTENTS

PREFACE

The mineral and gem hobby continues its amazing growth—especially in the United States and Canada, but also throughout many other parts of the world. Active rockhounds, mineral collectors, and amateur lapidaries are numbered in the hundreds of thousands in America alone. Millions more are interested enough in these activities, of which they have seen and heard so much, to justify this book.

Its purpose is to define, explain, and describe the most essential subjects in this fascinating realm of science and nature. In this handy-size book are also included important and interesting names and terms that you will need when learning about minerals, crystals, gems, the art of gem cutting, rocks, mining, meteorites, and the basic mineralogy and geology that underlie the entire field. Most of the subjects follow an alphabetic, encyclopedic style, and each is a concise definition. The minerals and gems are described individually; and in this section is included a selection of the curious and obsolete names often encountered in the literature. Related material is treated differently where desirable; there are short essays on mineral collecting, identifying minerals, the lapidary art, and the origin of mineral names, and information about the classification of meteorites and falls of unusual interest.

The superb color photographs by Reo N. Pickens, Jr., are among the finest ever made. They have been selected to show the features of the mineral kingdom in such a way as to inform and instruct, as well as to delight. The drawings, made by Christopher D. Wadsworth and Mignon Wardell Pearl, have been carefully chosen for their educational value. So too have the black-and-white photographs, credit for which appears in the acknowledgments. The author's thanks are also due to Mrs. J. V. Winters for assistance and to Chester H. Johnson, a former student of the author's and now Science Editor of Western Printing and Lithographing Company, for generous encouragement and cooperation.

GEMS MINERALS CRYSTALS *and* ORES

MINERAL COLLECTING

The value of a mineral collection depends largely on the quality of the specimens and on the completeness of the information concerning them. A bruised or scratched specimen, or a broken crystal, has little to recommend it, and a mineral of unknown locality or undefined associations can be admitted to a cabinet only under protest. The geographical and geological distribution is an important fact about any given species; it is therefore desirable, in each locality, to collect every variety that can be found. Specimens that would be worth little by themselves become valuable when studied in relation to others; and a common mineral, found under unusual circumstances, may have exceptional interest. "Waif" rock specimens, their exact sources unknown, are almost valueless.

The best place to look for minerals is always one that has been well opened, such as a mine, a quarry, road or railroad cut, or a cliff. Fresh exposures of unweathered rock yield the best specimens. However, in prospecting where no work has been previously done, it is desirable to get below the weathered rock surface to the fresh material beneath. Sometimes nature assists in this direction, as in ravines and along the bases of cliffs where, in early spring, rock falls off and leaves clean exposures. The recent talus at the foot of a cliff affords a good spot for examination.

The tools used in collecting minerals are few in number. The first and most indispensable tool is a hammer. It should be of well-tempered steel, weighing about two pounds, with the striking face square and the cutting edge parallel to the handle.

It often happens that specimens brought in from the field have an unnecessary amount of waste rock adhering, or part of a crystal or crystals may be covered up by the matrix. This can be

Portable rock saw for use in cutting samples or bulk specimens in the field.

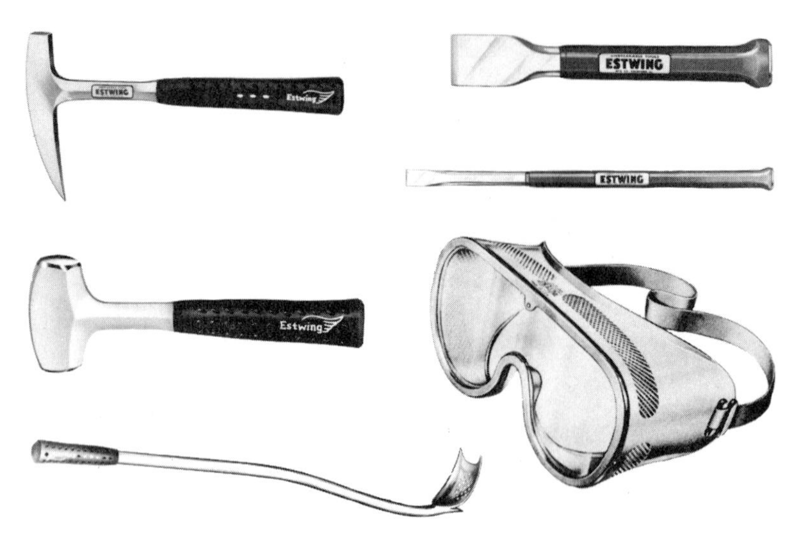

Rockhunting tools. A complete assortment of specimen-collecting implements includes: rock pick hammer, crack hammer, chisels, long-handled gem scoop, eye shield. The rock pick hammer is the most essential tool for the collector when in the field.

readily removed and the size of the specimen reduced by means of a few small chisels and cleaning tools and a trimming hammer. The trimming hammer should weigh from a quarter to half a pound, and have both faces square. By holding the specimen firmly in the left hand and delivering a sharp blow with the hammer, the waste rock may be removed without injury to the mineral. A pitching tool, or broad chisel, is useful in shaping the specimen after trimming. The other tools that can be useful are three steel chisels for cutting purposes, one 6 inches long, the others 3 inches long and of similar patterns; a set of steel wedges for splitting rocks; and a pickax to remove surface material and for prying.

The operation of blasting, from the standpoint of the mineral collector, requires the greatest care and judgment. A fine locality may be ruined by the reckless use of powder. Unless you are experienced in blasting, it is better to let someone else do it.

In collecting minerals, the labels should be written at once and wrapped with the specimens. The locality should be described as completely as possible. Unidentified minerals should be marked with full information as to mode of occurrence, in order that they may be properly studied later. It is always wise to preserve your records in some form of field notebook. Trust

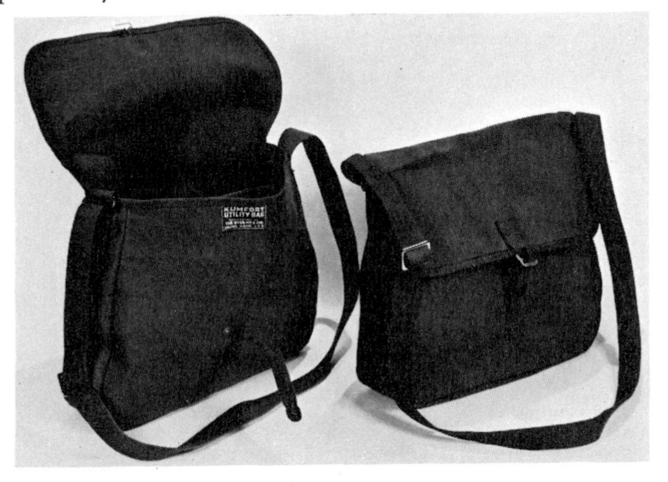

A carry-all, shoulder-strap knapsack for use in the field. Your tools, lunch, and specimens will be well accommodated in such a handy item. It's a must for the field collector, from pebble pup to scientist.

nothing to memory. The exact site of a rock body may be of considerable importance, especially when you want to visit it again.

For the transportation of the material to a place where it can be properly culled and packed, a knapsack or collecting bag is convenient. A good size is about 12 inches long, 12 inches high, and 4 inches deep. It should have a flap over the top, extending fully 6 inches down the opposite side, and a strap and buckle to fasten it in position. The carrying strap should be 2 inches wide and long enough to go over the shoulder or around the neck.

A supply of paper, preferably newspaper, is necessary for wrapping specimens. Never attempt to carry a specimen unwrapped in the bag. The material collected must not be permitted to chafe or rub.

Crystallized and delicate specimens should each be wrapped first in tissue paper, then in raw cotton, and finally in newspaper. Very fragile specimens, such as needlelike cuprite or natrolite, should be further protected by being boxed separately.

For shipping, the minerals are best packed in comparatively small boxes; large boxes are too heavy for proper handling. A layer of excelsior, straw, hay, grass, Spanish moss, coarse shavings, or other packing material (never sawdust), is put about 1 inch

Four-drawer cabinet for storage of smaller mineral, crystal, and rock specimens. Many specimens can be stored neatly in a small space.

deep on the bottom; then the heaviest and most massive specimens are packed firmly and closely. Spaces between them should be filled with excelsior. Then add another layer of excelsior, a layer of specimens, and so on until the box is filled. Finally a layer of excelsior is put on top and the cover fitted tightly. The box must be absolutely full before fastening. Coarse, massive material, collected in bulk to be broken up and trimmed later, may be transported in boxes or barrels without wrapping or other special precautions.

Double-tiered display cabinets with sliding, locking doors, and glass walls. This showcase can turn the serious collector's private collection into a museum exhibit at home.

Glass-walled showcase with sliding, locking doors provides an attractive display area and protection for choice mineral specimens. Such a case permits a full view of its contents from all angles.

GEM CUTTING

The lapidary trade—gem or stone cutting—began several thousand years ago, when minerals and rocks were first shaped and polished for ornamental purposes as objects of commerce. Much earlier than this, however, later Stone Age man had traded tools, ornamental stones, crystals, and shells over wide areas. Stones were fashioned for personal adornment on an individual basis, and artifacts—although designed for special uses—were worked generally in the same manner. This activity goes back a long way and, in fact, has given its name to man's stages of development. The Paleolithic and Neolithic periods in anthropology—the Old and New Stone Ages—are recognized as not only the oldest but also the longest aspect of human history. As a part of the building of homes, temples, and other structures, stone—aside from tool making—has long played a significant role.

The art and craft of the lapidary—as popular an activity as collecting—involves a diversity of equipment and methods. These range from the fashioning of rounded surfaces on the common quartz gems to the faceting of flat surfaces on more difficult gems, and then to the carving of elaborate ornamental objects from a wide range of materials. Some of the work is done more or less mechanically, with the aid of devices for making spheres or cutting and polishing facets. Even the simplest work, however, may require both care and skill. Mosaic and inlay work represent special though related techniques. Amateur efforts around the world often compare brilliantly with professional lapidary production, because time is taken by the amateur to experiment with ingenious ways to create new effects.

Except for the application of electricity to lapidary machinery and the introduction of synthetic abrasives (notably Carborundum), today's lapidary—amateur and professional alike—operates in basically much the same manner as did the medieval and ancient members of his craft. The raw material must first be extracted from the solid rock (unless it has been picked up loose), it must then be reduced to a convenient size and given an approximate shape, and afterward it is cut to final form, ending with the production of a satisfying polish.

The hobbyist is not likely to try diamond cutting at first, although it is not impossible that he may wish to later. Actually, the colored gems offer a much wider range of interest; agate, which is the most generally used of all the gems, comes in an endless diversity of colors and patterns.

The first step in the preparation of rough material is sawing, including slabbing and trimming. This is most often accomplished with the diamond saw, in which the diamond-dust abrasive is firmly attached to the saw, either in a notched rim or the more costly fused or sintered rim. Before the introduction of the diamond saw, the mud saw was used, and it is still seen in some places. A mixture of clay and silicon carbide is fed to the rotating steel blade. The blade that does the cutting is mounted on a shaft that is provided with a pulley and belt to connect it with an electric motor, which usually drives the saw at a surface or rim speed of 2,000 to 3,000 feet per minute. The specimen is carried forward, usually mechanically, on a moving carriage in which it is firmly clamped. A cooling liquid is kept in a tank beneath the saw, which dips into it. Steel ribbons and wire saws are used for very large specimens.

After sawing, the next operation is grinding. This process shapes the sawed slabs with a silicon carbide wheel, which should not be either too hard or too soft. A coarse (100 to 120 grit) and a fine (220 grit) wheel can be used alternately, rotating wet at perhaps 4,000 to 6,000 surface-feet per minute.

Then comes lapping, which gives the final shape to the specimen. A lap is a horizontal disk, usually metal and commonly 16 inches across, rotating at 250 to 300 revolutions per minute. Lapping is done in several steps, perhaps using 220-grit, 400-grit, and 1,200-grit silicon carbide successively, each being applied wet.

The stone is next smoothed with a flexible sanding cloth, mounted on a vertical or horizontal disk, a drum, or a belt passing over two pulleys. The strong cloth is covered with silicon carbide, which is used in two or more grits. This operation can be performed either wet or dry and at surface speeds of 3,000 to 4,000 feet per minute.

The final stage in gem cutting is polishing. The stone is pressed carefully against a polishing buff, which may be either a wheel or a disk of various materials, including felt, leather, cloth, and others. The polishing agent is one of a number of kinds of powder which is brushed wet onto the buff.

The fundamental forms in gem cutting are cabochon-cut, facet-cut, and carved stones. Some may also be perforated by drilling. A cabochon has rounded surfaces, although the bottom may be flat. If carried to completion, a cabochon may end as a sphere—this is a popular lapidary product for which special sphere-shaping devices are sold. A faceted gem—best typified by the "brilliant" cut on diamonds—requires more skill or else the use of mechanical aids, which can readily be bought. Carving is art—Michelangelo was a lapidary, and so are the natives of many lands who turn out stone animals, idols, and other tourist items.

The most important thing to do in making a cabochon is to select the right orientation of the rough material. You will want to show in the finished stone the best color (as in an opal) or the most attractive pattern (as in an agate). If the mineral is fibrous, the fibers should be parallel to the base of the gem. Cracks and other flaws should be eliminated. An aluminum pencil can be used to outline the stone, which is then sawed to a "blank" that is at least 1/8 inch larger than the final dimensions desired. A further reduction is made with a coarse grinding wheel, leaving 1/16 inch in excess. Mounted on a dop stick, the stone is then ground into a series of beveled bands, and these become the completed, rounded gem surface by the processes named above—fine grinding, coarse sanding, fine sanding, and polishing.

Facet cutting follows the same steps, but the numerous flat surfaces are arranged in pairs, series, or opposing sets. The lapidary who can put on the facets by observation alone is a skilled artisan. Most amateurs find it convenient to use one of the mechanical faceting accessories on the market.

A popular lapidary technique that has spread greatly in recent years is called tumbling. Rounded stones of irregular shape (baroque) are mass produced at little expense. Seldom is first-quality material used for this purpose, but the better stones are attractive in all kinds of costume jewelry. The polish is not equal to that obtained by hand, but it is pleasing enough.

The equipment for tumbling is varied. Some of the rotating cylinders have a capacity of 1 pound, whereas others hold 100 pounds or more. Tumblers can be made at home or bought from lapidary dealers, who have them in circular, hexagonal, or octagonal shapes, the last two being most recommended. Many amateurs use rubber tires as tumblers. The movement of the tumbler is usually accomplished by resting it on rubber-covered rollers driven by a small electric motor; in the tumbling process, stones in the top portion of the load are caused to slide over the underlying stones. This friction, combined with the action of the abrasives, furnishes the necessary polishing effect.

Each batch should be selected according to hardness, so that the product will be uniform. The abrasive is added and changed in increasing degrees of fineness—typically 80, 220, 400 grit—the last stage calling for the addition of a polishing agent. A "carrier" is often employed, this being composed of such strange things as nut shells, fruit pits, and sawdust. The tumbler is stopped and cleaned between each change of grit; a complete tumbling, from rough to final polish, requires several days or more.

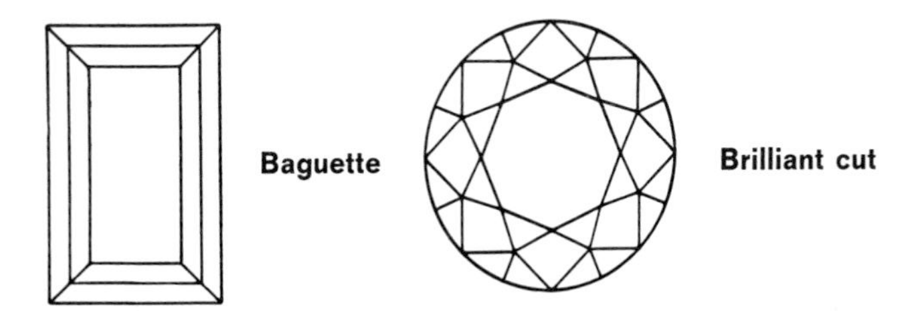

Baguette. A small gem cut in *step style*. Baguettes are usually set in a row or around a larger stone.

Baroque. Irregularly shaped. Oddly formed pearls and tumbled gems are described by this term.

Brilliant cut. A round style of gem cut typical of the diamond. The standard brilliant has 58 facets distributed as follows: 1 *table,* 8 *bezel facets,* 8 *star facets,* and 16 *upper-girdle facets* on the upper part (the *crown);* 8 *pavilion facets,* 16 *lower-girdle*

Brilliant-cut diamond

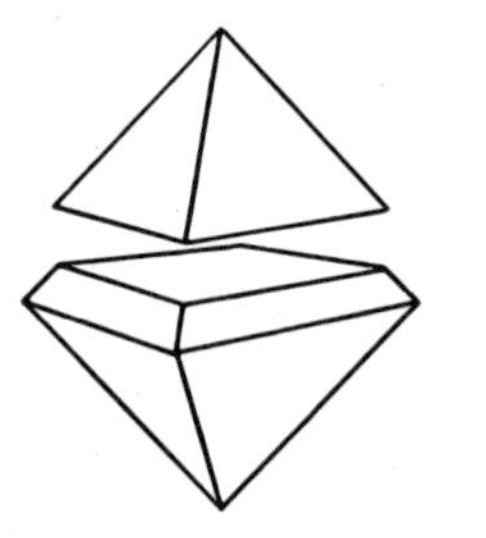

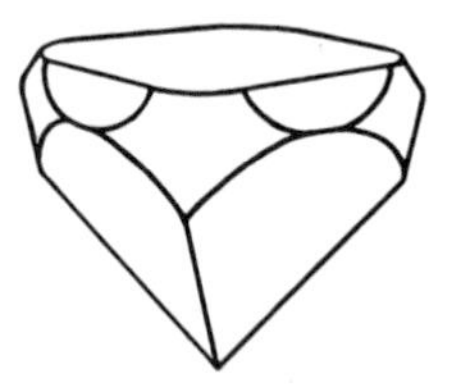

Shaping a brilliant-cut diamond from an octahedron

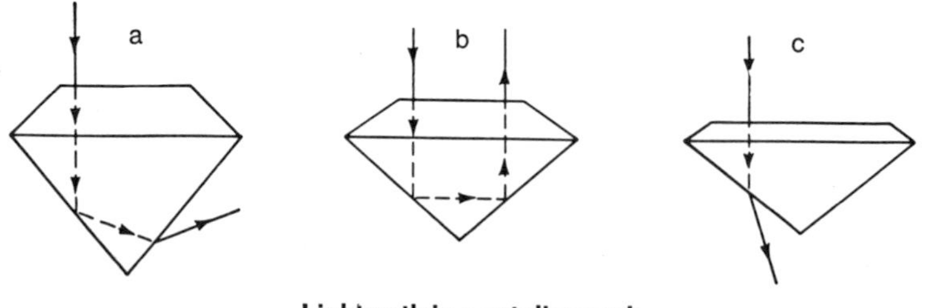

Light path in a cut diamond:
(a) too deep; (b) properly cut; (c) too shallow

facets, and 1 *culet* on the lower part (the *pavilion*). The proportions and number of facets vary according to the fashion and size of the stone. A number of so-called fancy cuts are modifications of the brilliant cut, yielding such cuts as *marquise* (boat shaped), *pear, pendeloque* (pointed-pear shaped), *heart shape,* and *half moon.* A simple style of circular cut, used mostly for small diamonds, is the *single cut,* having 17 or 18 facets.

Complete cabochon-making lapidary unit. On this versatile machine the amateur lapidary can turn out finished gem cabochons from rough stones. The wheels (from left to right) are grinder, sander, two polishing belts of coarse and finer grits, and a flat buffing wheel on the end.

Cabochon cuts

Cabochon. A style of cutting with a rounded top. This is the simplest type of lapidary work, although it is often used with difficult materials.

Carat. A unit of weight for gems. The standard metric carat is 200 milligrams. It is divided into 100 *points.*

Coolant. A lubricating and cooling agent used in lapidary work. It is usually oil or kerosene or both.

Crown. The upper part of a faceted gem, above the girdle. On the standard brilliant cut or modifications of it, the crown has 33 facets.

Culet. The small, bottom facet on a cut gem.

Cutting. The shaping of lapidary material. The products of gem cutting are divided into three styles: cabochon, faceted, and carved and engraved stones, according to the nature of the material and the effect desired. Diamond, alone of all substances, is cut and polished in the same operation.

Diamond-set blades for use in various lapidary machines for sawing and trimming rocks and minerals. These tough blades can cut many hundreds of stones before showing much wear.

Diamond saw. A lapidary saw set with diamonds. It is made of phosphor bronze or special steel. The diamonds *(bort)* are inserted into small notches around the rim, or else they are mounted into an attached rim made of powdered metal and diamond powder, which is heated until the metal fuses *(sinters).* For slabbing, the diamond saw has replaced the slower and dirtier mud saw.

Dop. A lapidary holder to aid in cutting and polishing stones. The dop stick is usually a rod of hardwood 4 to 6 inches long, on which the stone is gripped at one end in heated wax or (for cold dopping) liquid adhesive. Dops for mechanical faceting are made of steel and held in an aluminum transfer jig.

Emerald cut. The *step cut* adapted to emerald.

Facet. A flat surface cut on a gem. Faceted gems are more difficult to make than those cut in cabochon style, although the work

Portable lapidary unit with faceting head. Such precision equipment enables the advanced lapidary to cut accurately the better gems in various faceted styles. Note the various dops.

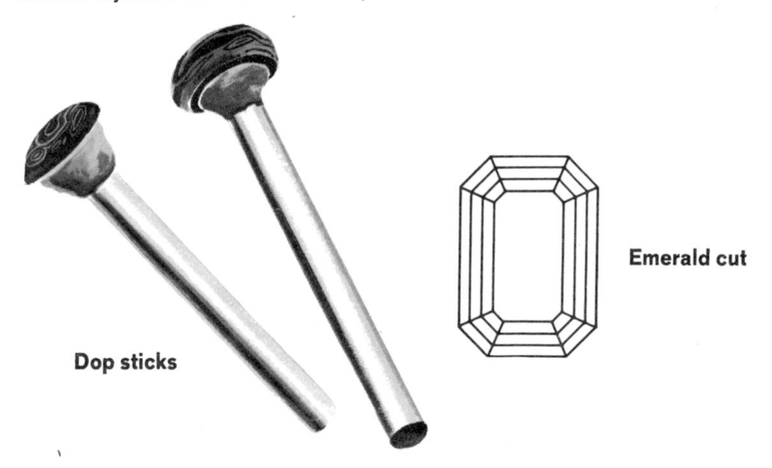

is simplified and standardized by the use of various mechanical devices, such as the facet head.

Fancy cut. Any style of gem cut other than the standard brilliant cut or step cut.

Girdle. The outer edge of a faceted gem. Sometimes it, too, is covered with facets.

Grain. A unit of weight for pearls. It equals ¼ carat.

Grinding. Shaping rounded lapidary material to its final form. This is often done in two steps, using a wet silicon-carbide wheel or disk for coarse work (100 grit) and another (220 grit) for fine work. A common size of wheel for gem cutting is 6 to 10 inches in diameter and ½ to 1½ inches in thickness, running at 1,530 to 3,820 revolutions per minute.

Lapidary. A person who cuts, polishes, and/or carves stones. Anything pertaining to gem cutting and the working of ornamental stones is included in this term, which is preferable to *lapidist* or other forms of the word.

Lapping. Shaping flat lapidary material to its final form. A disk of metal or other material, perhaps 16 inches in diameter, is rotated horizontally at 250 to 300 revolutions per minute. The wet abrasive is loose silicon carbide of various grit sizes (220 to 1,200), used in separate stages.

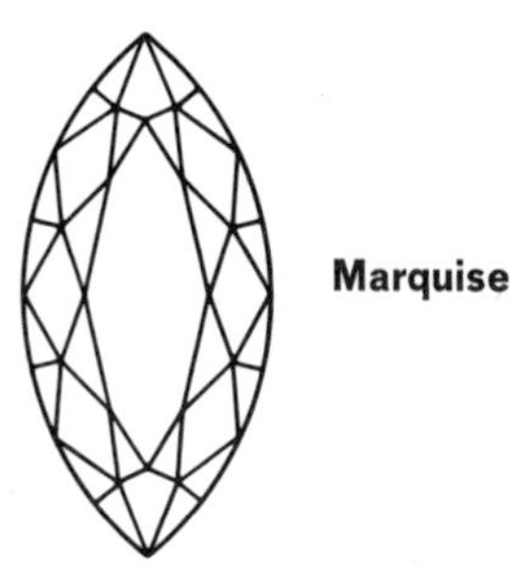

Marquise cut. A boat-shaped modification of the brilliant cut for gems. It is a popular fancy cut for diamond.

Mud saw. A lapidary saw using steel disks and a loose abrasive of silicon carbide and wet clay (the "mud"). It is used to cut sections or slabs of material for further treatment, especially trimming. The specimens are clamped in a vise and pulled against the mild-steel blade by weights. The blade, or series of parallel blades, rotates on an arbor in a covered box. In small sizes, the mud saw has been superseded by the cleaner and faster, but more expensive, diamond saw.

Pavilion. The lower part of a faceted gem, below the girdle.

On a standard brilliant cut or modifications of it, the pavilion has 25 facets.

Pendeloque cut. A pointed, pear-shaped modification of the brilliant cut for gems. It is a popular fancy cut for large, colored, transparent stones.

Point. A unit of weight for gems. It is $\frac{1}{100}$ of a carat, or 2 milligrams.

Vibrating lapidary machine for grinding and polishing. Rapid vibration (with various abrasives) abrades, smooths, and polishes.

Lapidary machine for polishing flat surfaces. This machine is especially adapted for polishing large pieces, such as bookends, paper weights, and museum exhibits. Horizontal rotation polishes one surface (the top) at a time.

Polishing. Putting the final gloss on lapidary material. The buff is made of leather, felt, muslin, canvas, or wood. A paste made of tripoli, cerium oxide, tin oxide, aluminum oxide, or other metallic oxides (according to the material being polished) is used as the abrasive. Diamond is the only substance on which cutting and polishing are done in a single operation.

Preform. A commercially shaped blank of gem material for lapidary work. Its use saves the preliminary operations of cutting.

Sanding. Smoothing lapidary material coming from the grinding wheel, before polishing it. Rough spots and grinding marks

Portable, table-model lapidary outfit—complete with motor, trim saw, grinding, sanding, and polishing wheels, and supplies for polishing. Many a fine piece can be turned out on this adaptable equipment.

are removed by sanding cloth made with silicon carbide. This is mounted on a drum, disk, or belt, 4 to 10 inches in diameter and running at 1,145 to 3,820 revolutions per minute. Two grits (120, 220), or more, of abrasive cloth are usually used. Sanding is done either wet or dry.

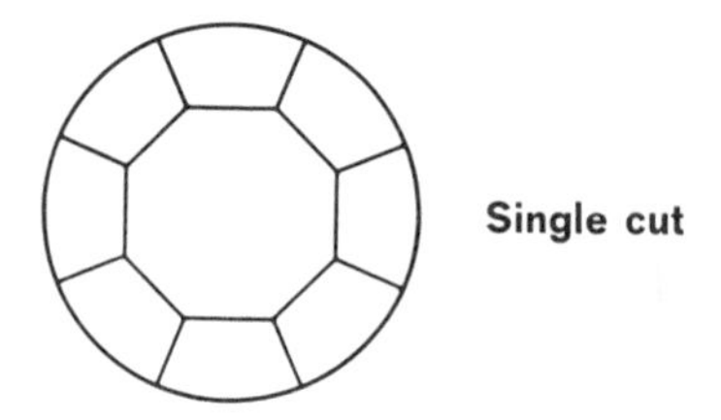

Single cut. A round style of gem cut. It has 17 or 18 facets (with or without a culet) and is used mainly for diamonds too small to warrant using the full brilliant cut.

Slabbing. Preparing cut sections of lapidary material for further treatment. The slices are generally $\frac{3}{16}$ to $\frac{1}{4}$ inch thick. The average speed of the slabbing saw is 850 to 1,000 revolutions per minute. For all but the largest sizes, the diamond saw has largely replaced the mud saw.

Step cut. A rectangular style of gem cut. The rows of facets

Large slab saw for cutting rough bulk specimens into slabs of varying thicknesses previous to trimming. A slab saw may be the first machine used in the lapidary process.

Lapidary's polishing or buffing machine. Pre-sliced and trimmed gem materials, called preforms, can be turned into polished cabochons on this type of equipment.

are arranged in steps, usually in sets of three, parallel to the girdle. This cut is especially popular as the *emerald cut* for emeralds, and it is fashionable for diamonds and suitable for any colored gem. Modifications, as *fancy cuts,* include *kite, triangle,* and *keystone.* Small, step-cut gems are also called *baguettes.*

Table. The large, top facet on a cut gem.

Templates

Template. An outline of gem sizes and shapes, like a stencil. Through the variously shaped holes are marked the part of the lapidary material to be saved, the pattern of the finished stone.

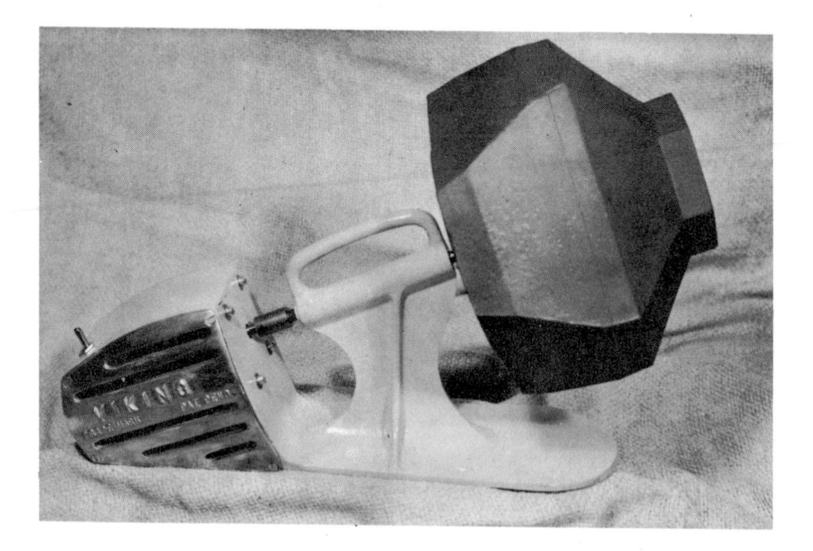

Two models of tumblers—the upper rotates, the lower vibrates. Polished gem stones of irregular shapes and sizes, known as baroques, can be turned out within a few days in considerable quantity and with little effort.

Lapidary's trim saw. Connected to an electric motor, this tool uses a diamond-set saw blade to trim and give general shape to a specimen before sanding and polishing.

Trimming. Reducing sections of lapidary material coming from the slabbing saw. The diamond-set, steel trim saw is usually 6 to 10 inches in diameter and rotates at 1,725 revolutions per minute; smaller and thinner blades of phosphor bronze are run faster for slitting valuable material. The trim saw can also be used for various hollowing and carving operations.

Tumbling. Producing irregularly rounded gems (as mentioned in the previous paragraphs) by rotating or vibrating them with various wet abrasives in motor-driven drums. Sometimes such things as sawdust or nut shells are added to carry the cutting powder.

IDENTIFYING MINERALS

There are many ways to identify minerals. The procedure and techniques you should use will depend upon the problems presented by a particular mineral, upon the available equipment—and most important of all, upon your degree of skill.

Some minerals can be identified at sight, requiring only a familiarity with their appearance (crystal form or habit) and their other properties (such as color, luster, and cleavage) that can be determined by observation alone. For this reason, you should take every opportunity to examine labeled specimens in museums and private collections.

A still larger number of minerals can be identified by easy tests, either in the field or the laboratory. It is usually more desirable to rely upon the simpler physical properties for field testing, because it is rather a nuisance to attempt blowpipe and chemical tests outdoors. When the specimens have been brought home, the more detailed observations can then be made.

Key properties, such as luster, color, streak, cleavage, fracture, hardness, magnetism, and specific gravity, can be used for mineral identification in the field. The only items of equipment needed are such everyday things as a streak plate, a knife blade, a magnet, and a means of finding the relative heaviness of a specimen—which can, with experience, often be estimated merely by lifting it.

The next step in identifying minerals is with a flame, using a blowpipe and the other inexpensive materials associated with this procedure.

The relative ease with which a mineral fuses when heated may be determined in most cases by comparison with the standard minerals of the Von Kobell scale of fusibility.

To make flame tests, some elements can be volatilized only by

NATIVE ELEMENTS

Arborescent silver on calcite. From Keweenaw County, Michigan, comes this superb example of metal "sculpture." The bright luster is still present, but long exposure to the atmosphere will tarnish it.

Native silver and native copper.

This grotesque form characterizes the crystallization of the metals. This pair is representative of the copper country of northern Michigan.

Native copper on calcite.

The bright, copper-red meshwork of twisted, distorted filaments is typical of this element in its native form. This fine specimen comes from northern Michigan.

Sulfur crystals.
These lustrous crystals come from Sicily, where they are found in exceptional sizes and quality.

SULFIDES

Millerite enclosed within calcite. The hairlike, radiating crystals of millerite are typical of this nickel sulfide. From Hamilton, Illinois.

Oxide-coated galena. The familiar lead-gray color and luster of galena is disguised, except at the very base of the specimen, by the iron-oxide coating. This large group is from Galena, Illinois.

Marcasite on calcite. The calcite was formed before the marcasite became sprinkled upon the crystal tips. From Mexico.

Sphalerite. Also called *ruby zinc* because of its similarity in color to the precious gem. From Baxter Springs, Kansas.

Octahedral galena.

This important lead-sulfide mineral is not often seen in this crystal form. So fine a specimen from Baxter Springs, Kansas, is of unusual interest.

Marcasite dollar. With its bright center and wide, radiating rim, this disk illustrates an unusual crystal habit. From Joplin, Missouri.

Marcasite on iridescent sphalerite.

Two well-known and commonly associated sulfide minerals. A product of the Tri-State district, where the corners of Missouri, Kansas, and Oklahoma adjoin.

Marcasite. This is a perfect example of *cockscomb marcasite;* note the crested effect of the radiating crystals. From Baxter Springs, Kansas.

Cubic pyrite. Satellite cubes, showing the striations well, are clustered upon the main crystal. This fine specimen is from Arizona.

Galena and sphalerite crystals on chert. Note the galena cube. The sphalerite is the ruby-zinc variety. From the Tri-State district.

Molybdenite. This specimen displays a foliated habit. The lead-gray color and metallic luster are characteristic. From Washington.

Pyrite. These magnificent crystals from the classic locality of Leadville, Colorado, are mostly pyritohedrons, a crystal form that derives its name from the mineral. Note the characteristic striations.

Galena crystals on marcasite. These two sulfide minerals often occur together, as in this combination from Galena, Illinois.

SULFO SALTS

Boulangerite.

These bluish lead-gray crystals are from Washington.

Jamesonite. This delicate specimen is from Mexico. The fine needles are brittle crystals.

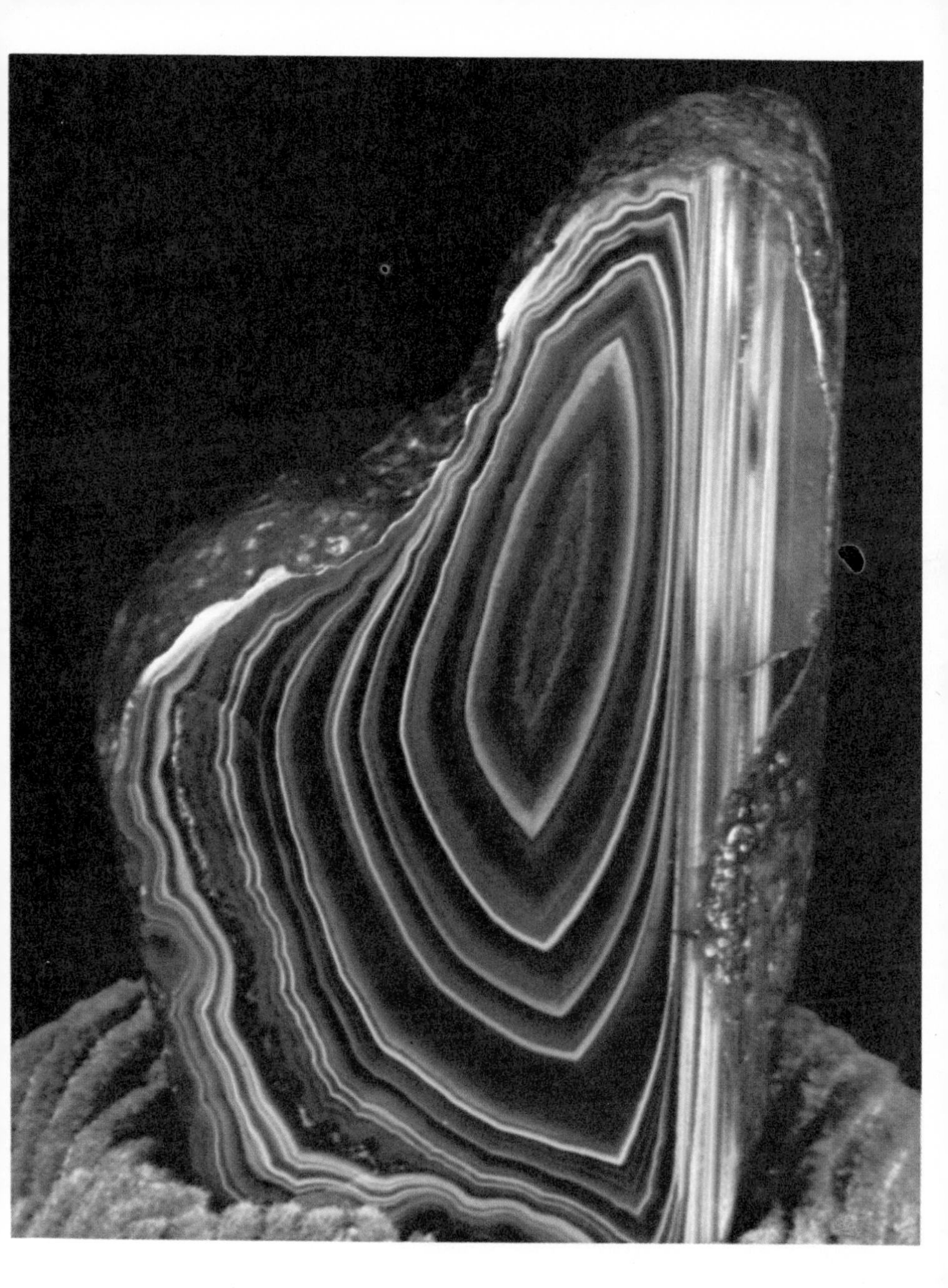

OXIDES

Agate. The rich coloring and the distinctive and complete pattern of this Lake Superior agate from Winona, Minnesota, make it a museum piece. The straight banding on the side represents the original base of the agate.

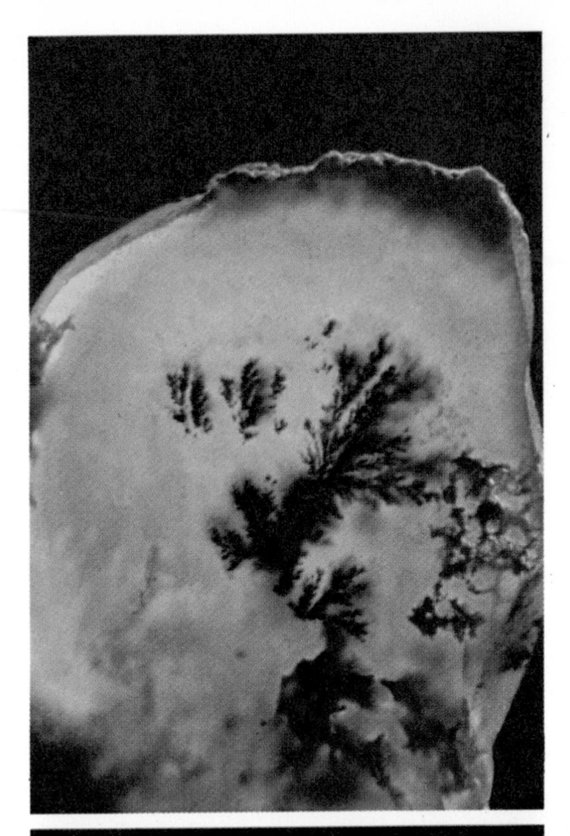

Moss agate.

So named for the branching pattern of the pyrolusite inclusions. From Oregon.

Bull's-eye agate.

The name of this agate from Winona, Minnesota, is most appropriate.

Opal in seam. This example of *black opal,* as distinguished from the whiter type, comes from Lightning Ridge, Australia. Opal's wonderful play of colors sets it apart from any other mineral.

Quartz crystals. These crystals start with a single base but terminate separately. From Hot Springs, Arkansas.

Botryoidal chalcedony geode. A study in petrified bubbles, from Hamilton, Illinois. The milky-blue color is typical of chalcedony, and so is the botryoidal growth.

Cassiterite.

This especially handsome aggregation is from one of the world's important tin-producing mines in Bolivia.

Quartz intergrowth.

This intricate specimen from Hot Springs, Arkansas, shows the varying directions of growth in one group.

Quartz.

Like a burning torch or a jeweled tiara are these beautiful crystals from Bisbee, Arizona. Their redness is due to inclusions of hematite.

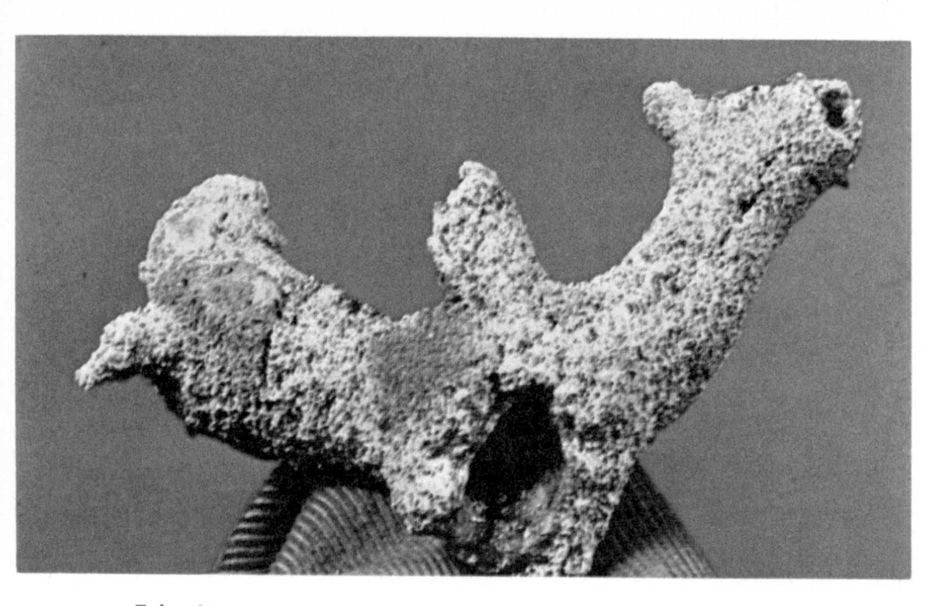

Fulgerite. This antlerlike contortion is the result of the fusion of sand when struck by lightning. The inside of a fulgerite is typically a glassy tube. This example is from Michigan.

Chalcotrichite. In its intricate network of very fine, spiny, red crystals, this mineral is the needle form of cuprite. This colorful specimen is from Bisbee, Arizona.

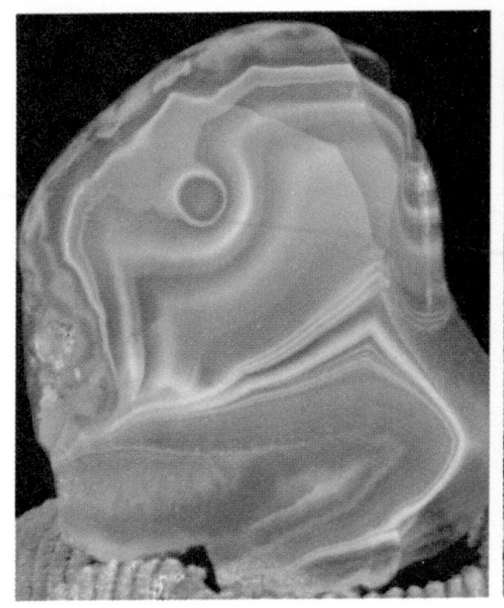

Agate. This Lake Superior agate, from Winona, Minnesota, looks like an abstract painting or a frog on a rock.

Pyrolusite. Sheafs of bluish-steel needles, jutting out from their matrix, dominate this view of the crystals. From Creede, Colorado.

Quartz crystals with inclusions of hematite. Hematite causes the brownish-red cast. An interesting specimen from Canada.

Smoky quartz topped by limonite pseudomorph after siderite. Note the strangely human, mineralized "face." The quartz has been darkened by radioactivity. Crystal Peak, Colorado.

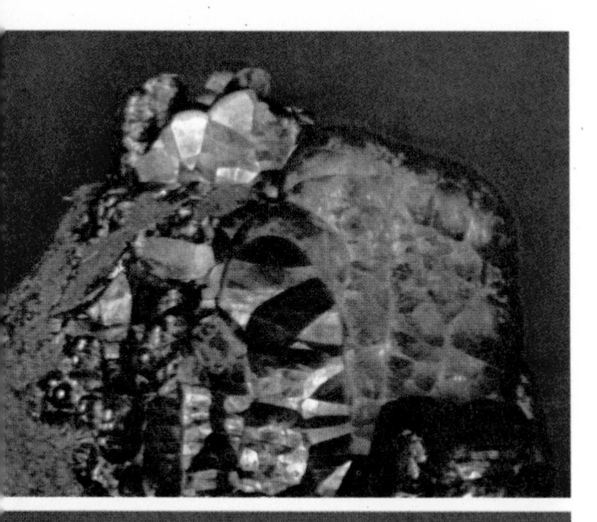

Botryoidal hematite.

This variety of hematite is also called *kidney ore* from the rounded shape of its surface. From Ishpeming, Michigan.

Amethyst.

This lavender crown of amethyst crystals from Canada would grace any mineral collection. Tiny crystals of a metallic mineral, perhaps pyrite or galena, dot some of the faces of the amethyst.

Quartz penetration.

It looks as if a glass spike had been driven into this specimen. From Hot Springs, Arkansas.

Chrysoberyl twin.

Twinning is the usual crystal habit of this gem mineral, and the asparagus-green color is common. Madagascar, source of many of the world's finest minerals, was the home of this specimen.

Pseudobrookite.

Nestled in a cloud of quartz crystals, this little burst of black, bladed crystals stands out in sharp contrast. This specimen is from Utah.

Uraninite crystal.

This is of special interest, as the mineral is a primary source of radium. From Cardiff, Ontario.

CONTINUED ON PAGE 113

fusing the powdered mineral with a flux, such as borax or sodium carbonate, or by moistening the mineral or powdered mineral with hydrochloric (muriatic) acid or sulfuric acid.

Certain elements can be identified by color when dissolved in a flux, to which the element's characteristic color is given. Borax is most commonly used. Sodium metaphosphate and microcosmic salt (sodium ammonium hydrogen phosphate) are called "salt of phosphorus." Sodium carbonate and sodium fluoride are also used as fluxes for identifying a few elements. The flux is made into a bead by heating it on the loop of a platinum wire. A small amount of the powdered mineral is picked up on the bead and fused in the flame of a blowpipe, first in the oxidizing part, then in the reducing part. Metallic minerals should first be roasted on charcoal in the oxidizing part of a blowpipe flame or in an open tube, before being taken up on the bead, in order to obtain an oxide and remove troublesome elements, such as arsenic. The color reaction depends upon the flux used, the amount of mineral, the presence of other elements, the type of flame (whether oxidizing or reducing), and the temperature (whether hot or cold).

Some elements in minerals can be determined by heating the powdered mineral in a glass tube that is open at both ends. These are known as *open-tube tests.* Various oxidized products are obtained, and these are identified mainly by their color, odor, and distance of condensation from the heated sample.

Some elements in minerals can be determined by heating the powdered mineral in a glass tube that is open at one end and closed at the other. These are known as *closed-tube tests.*

Many constituent elements of minerals can be determined by heating the powdered mineral on a block of charcoal. The charcoal serves as a support and as a reducing agent. Either reduction or oxidation may be made on charcoal, according to the part of the blowpipe flame (reducing or oxidizing) that is used, and the fluxes that are added. The results obtained include oxide, iodide, and cobalt nitrate coatings, and metallic globules.

In addition to observing the ease of fusibility, magnetism, decrepitation (crackling when heated), fluorescence, phosphorescence, and glowing, you will find other tests useful. These involve changes of color, emission of gases, and techniques for the deposition of sublimates (coatings).

Chemical tests are perhaps the next step in mineral identification. These range from simple ones, such as the effervescence (fizzing) of a carbonate mineral in acid, to complex ones that need the services of a good laboratory. Many of these useful tests can be made by the hobbyist.

Still other minerals require more specialized means of identification. The petrographic microscope is an astonishingly effective tool for this purpose; besides a magnified view of the structure and cleavage, it reveals the index of refraction and other optical characteristics of the mineral. The spectroscope is often used to determine the chemical composition of minerals; it is especially useful when certain rare elements not otherwise easy to identify are present.

The ultimate in mineral identification is probably by the use of "powder photographs." This X-ray technique yields a strip of film upon which appear concentric arcs, which are parts of concentric circles, which in turn are parts of concentric spheres. Measuring the distance between these arcs and listing them by their relative intensity enables the operator, by referring to published data, to name the mineral, perhaps after a quick check against some physical property or chemical constituent.

New methods for identifying minerals are being devised each year, and changes and improvements can be expected in the future. Some of the present methods are highly complex, each being adapted to special problems.

MINERAL TERMS

Absorption. Removal of some of the colors or wave lengths from light passing through a mineral or gem. The color that is seen is a combination of the colors or wave lengths that are not absorbed by the mineral or gem.

Acicular. A mineral habit consisting of needlelike crystals. Rutile is a good example, as in rutilated quartz, in which the slender needles of rutile penetrate the quartz. Stibnite, natrolite, and pectolite also occur in acicular crystals.

Adamantine. The hard, brilliant luster of diamond. Transparent lead minerals, such as cerussite and anglesite, are good examples.

Agatized wood. Petrified wood in which the filling and replacing silica has the gemmy form of agate.

Alkali. A soluble salt found on or near the surface in arid regions. It is usually sodium carbonate or potassium carbonate. Because of an analogous composition, an *alkali feldspar* contains much potassium or sodium and may be orthoclase, microcline, anorthoclase, or albite, as well as combinations, such as perthite.

Alloy. A mixture of a metal and another metal or nonmetal. Brass is an example of an alloy of two metals—copper and zinc. Steel is an example of an alloy of a metal and nonmetal—iron and carbon.

Alumina. Aluminum oxide, Al_2O_3. Corundum is composed solely of alumina.

Aluminate. A multiple oxide mineral containing aluminum and another metal atom bonded equally to oxygen. Spinel and chrysoberyl are important examples.

Amorphous. Lacking crystalline structure. Mineraloids are amorphous, and so are glass and liquids.

Amygdaloidal. A mineral habit consisting of an almond-shaped growth. When it fills a cavity *(vesicle)*, it is termed an *amygdale* (or *amygdule),* such as the quartz, calcite, and zeolites that are common as cavity fillings in lava.

Antimonate. A mineral composed of antimony, oxygen, and other elements. Bindheimite is an example.

Antimonide. A mineral composed of antimony combined with a more metallic element or group of elements. Dyscrasite and breithauptite are among the few antimonide minerals. Ullmannite and gudmandite are sulfantimonides or antimony-sulfides.

Apache tear. A rounded lump of obsidian. Of various sizes, these pellets of black glass are the cores of the shell-like structures occurring in perlite (a glassy, igneous rock filled with concentric, spheroidal cracks).

Argentiferous. Silver bearing.

Arsenate. A mineral containing the arsenate ion (AsO_4). Important examples of this class include erythrite, scorodite, and mimetite.

Arsenide. A mineral composed of arsenic combined with a more metallic element or group of elements. Niccolite and skutterudite are examples of this chemical class of minerals. Cobaltite and arsenopyrite are sulfarsenides or arsenide-sulfides.

Assay. The analysis of gold and silver ore to determine its precious-metal content. This is *fire assaying,* which uses a furnace and involves the processes of *scorification* (separation by heating with a lead and borax flux) and *cupellation* (melting and oxidizing the lead).

Star stone showing asterism

Asterism. A starlike pattern of light seen in certain minerals. It is caused by the scattering effect of inclusions, hollows, or concentrations of coloring matter. Asterism is shown by the star gems, such as star ruby, star sapphire, star garnet, and star quartz, and by phlogopite. The number of rays depends on the crystal system of the mineral and the direction in which the stone is cut.

Astringent. A taste that puckers the mouth. Alum has an astringent taste.

Atom. The smallest particle having the properties of an element.

Auriferous. Gold bearing.

Bladed. A mineral habit consisting of flat crystals shaped like a knife blade. Kyanite shows it well.

Blowpipe in use

Blowpipe. A metal tube for concentrating heat and directing it onto a mineral. The operator blows a narrow stream of air against a small flame in a candle, alcohol lamp, or Bunsen burner. The specimen or powder is held in forceps, on a block of charcoal or plaster of paris, in a loop of platinum wire, or in open or closed glass tubes. Mineral identification by blowpipe methods includes the production of coatings (sublimates) and metallic globules on the charcoal or plaster, flame and bead tests on the platinum wire, and chemical tests in the glass tubes.

Borate. A mineral containing the borate ion (BO_3). Among the more common borate minerals are kernite, tincalconite, borax, ulexite, colemanite, howlite, and boracite.

Botryoidal. A mineral habit consisting of a rounded group looking like a bunch of grapes. Hematite, psilomelane, malachite, and prehnite show it. Clusters of chalcedony with this habit are popularly called *grapestone.*

Bromide. A mineral composed of bromine and another element or group of elements. Bromyrite is an example.

Calcic. Containing calcium, Ca.

Capillary. A mineral habit consisting of flexible, hairlike crystals. Examples are millerite and the chalcotrichite variety of cuprite.

Carbonaceous. Containing organic carbon. Coal, jet, and amber are carbonaceous in composition.

Carbonate. A mineral containing the carbonate ion (CO_3). All carbonates dissolve and effervesce in acid, releasing carbon dioxide gas; for some, you may have to powder the mineral or heat the acid. The more important carbonate minerals are conveniently divided into the *calcite group* (calcite, magnesite, siderite, rhodochrosite, smithsonite), *aragonite group* (aragonite, witherite, strontianite, cerussite), *dolomite group* (dolomite, ankerite), and *copper group* (malachite, azurite).

Cave marble. Banded calcite or aragonite in caverns. It is also called *cave onyx,* but it is neither true onyx nor marble.

Cave pearl. A smooth, round deposit of calcite or aragonite deposited around a nucleus by dripping water in a cavern.

Cavestone. Mineral deposits in caverns, including stalactites, stalagmites, columns, pillars, helictites, cave pearls, and other forms of calcite, aragonite, and other minerals.

Chatoyancy. A moving beam of light seen in certain minerals when they are turned. It is caused by fibers or other inclusions or hollow canals arranged at right angles to the beam. Chatoyant gems, such as cat's-eye and tiger's-eye, are valued for this attractive effect.

Chemical composition. The chemistry of a mineral. The chemcal composition is expressed in a shorthand notation known as a *chemical formula.* The elements are named by a symbol which is a one-letter or two-letter abbreviation of the English or Latin word for the element, as Ca for calcium and Au for aurum (gold). The number of atoms of the element in a compound is stated by the subscript number following the symbol, as SiO_2 (one atom of silicon and two of oxygen) or Al_2O_3 (two atoms of aluminum and three of oxygen).

Chloride. A mineral composed of chlorine and another element or group of elements. Halite, sylvite, cerargyrite, atacamite, and carnallite are examples.

Chromate. A mineral containing the chromate ion (CrO_4). Of the few chromate minerals known, crocoite is the most important.

Class. A major division of minerals, based on chemical composition. The classes may be designated in several ways. The usual ones include: ELEMENTS, SULFIDES, SELENIDES, TELLURIDES, ARSENIDES, SULFOSALTS, OXIDES, HYDROXIDES, HALIDES, CARBONATES, NITRATES, BORATES, SULFATES, CHROMATES, PHOSPHATES, VANADATES, ARSENATES, TUNGSTATES, MOLYBDATES, URANATES, SILICATES. Classes are subdivided into *families* or *types*.

Cleavable mass. The occurrence of a mineral in irregular pieces showing good cleavage throughout.

Cubic cleavage in halite

Cleavage. The splitting of a mineral in a definite crystal direction. It is an important way to identify minerals. The flat surface that is produced usually has a pearly luster, which distinguishes it from a natural crystal face. Cleavage is described by its quality and direction, such as "perfect octahedral" for diamond. Cleavage is always parallel to an actual or possible crystal face. Rocks also show cleavage, but this (also called *fissility*) is merely a tendency to break readily and has no definite meaning, as it does in crystallography.

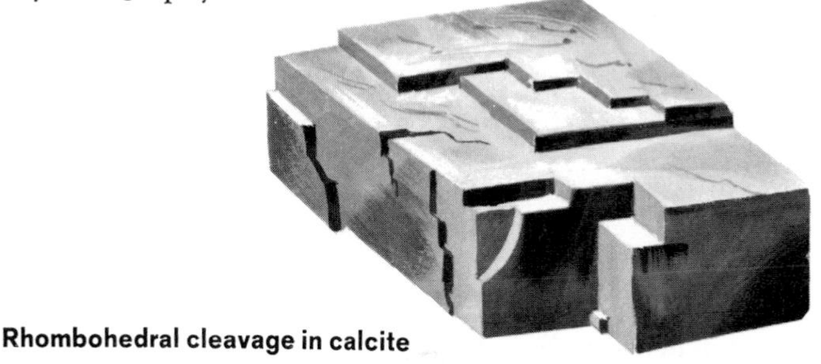

Rhombohedral cleavage in calcite

Cohesion. The ability of a mineral to hold together.

Columbate. A multiple oxide mineral containing niobium and another metal atom bonded equally to oxygen. Columbite, fergusonite, and pyrochlore are examples.

Columnar. A mineral habit consisting of individual crystals having a pillarlike shape. Tourmaline sometimes shows this habit very clearly.

Compact. A mineral habit of fine-grained, firm material. Alabaster is a compact variety of gypsum, capable of being carved with a knife.

Concentric. A mineral habit consisting of layers or bands surrounding a common center. Stalactites grow in a concentric pattern, as do agates, oölites, and concretions.

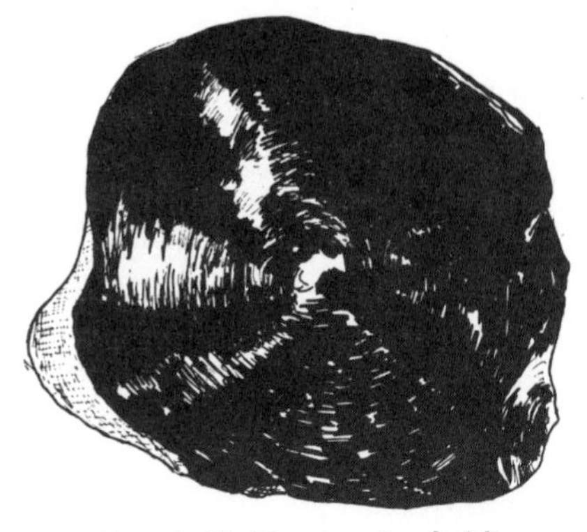

Conchoidal fracture in obsidian

Conchoidal. A fracture pattern in minerals and rocks, resembling the concentric arcs seen on shells. Glass and obsidian show it most typically, and quartz often does.

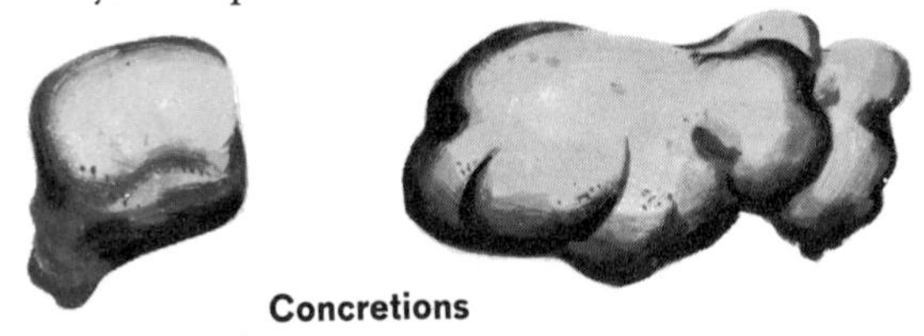

Concretions

Concretion. A nodule, or lump, of mineral origin. It forms around a nucleus consisting of sand or a fossil or other material. Silica, calcium carbonate, iron sulfide, and other chemical compounds are often represented by concretions.

Crested. A mineral habit consisting of groups of tabular crystals arranged in ridges. Marcasite is an example of a crested mineral.

Dana system. The mineral classification most used in North America. It is based upon crystal chemistry and combines both the sciences of crystallography and chemistry in a systematic way, as determined by X-ray studies and chemical analyses. The minerals are also numbered, beginning with native gold and extending through the various chemical classes and ending with the silicates.

Dendritic. A mineral habit consisting of branchlike crystals in contact with one another. Pyrolusite forms dendritic crystals in moss agate and on slabs of limestone and other rock. Native gold, silver, and copper are sometimes dendritic and may then make very handsome specimens.

Density. Specific gravity. The two terms are technically distinct but are expressed by the same numbers, which indicate the weight of a mineral as compared with the weight of a similar volume of water.

Diaphaneity. Degree of transparency. Minerals are described as *transparent* (entirely clear), *translucent* (light comes through thin edges), and *opaque* (transmits no light).

Dichroism. Showing different colors when viewed in two different directions. It is seen in crystalline, colored gems that do not belong to the isometric system. With a dichroscope, a number of gems can be identified by this property. For example, ruby can be distinguished from garnet, sapphire from spinel, and emerald from tourmaline. When more than two colors are involved, the general term *pleochroism* is used.

Dimorphism. A chemical compound crystallizing as two different minerals. Thus, pyrite and marcasite, both being iron sulfide (FeS_2), are dimorphous.

Double refraction. The separation of a ray of light into two

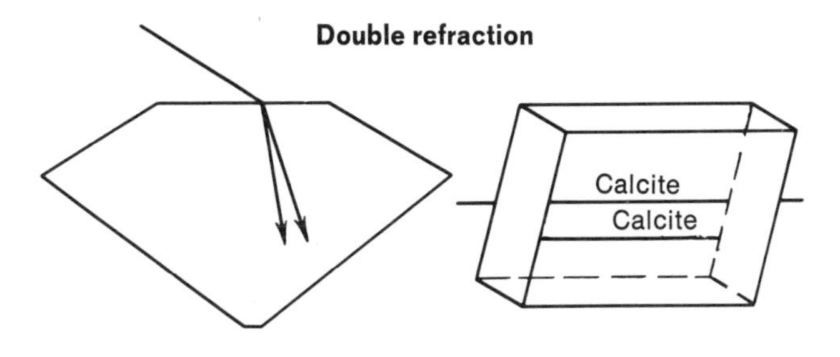

rays. Each may have its own index of refraction. Because of this property, calcite (Iceland spar) shows a doubling of an image seen through it.

Drusy. A mineral habit consisting of a growth of small crystals on a surface or in a cavity. Quartz is often drusy, making sparkling specimens.

Ductility. The ability of a mineral to be drawn into a wire. Native copper is extremely ductile.

Earthy. A mineral habit consisting of fine-grained, powdery material. Limonite often appears earthy.

Also, a luster that is dull. Clay minerals have an earthy luster.

Effervescence. The property of a mineral or rock whereby it bubbles in acid. The carbonate minerals and rocks fizz in acid, releasing carbon dioxide gas (CO_2).

Efflorescence. A powdery crust formed by evaporation from the surface of a mineral or rock. Borax is an example of a mineral that effloresces in dry air. Alkali is a result of efflorescence.

Elasticity. The ability of a mineral to be bent and then to regain its original shape when the pressure is released. This is the chief distinguishing property of the mica group of minerals, all of which are elastic.

Element. One of the distinctive chemical substances of which rocks and minerals are composed. There are presumably 92 natural elements in the known universe, and others have been made artificially in recent years.

Etched. Having the surface corroded by chemical action. This can take place either naturally or artificially. Etch figures are useful in identifying certain minerals and determining their atomic structure.

Family. A subdivision of a *mineral class,* based on a decreasing proportion of positive to negative ions or electrostatically charged atoms. Family is the same as *type.* Mineral families may be further subdivided into *groups.*

Felted. A mineral habit consisting of closely matted fibers. Asbestos is often felted.

Ferrate. A multiple oxide mineral containing iron and another metal atom bonded equally to oxygen. Magnetite, franklinite, and chromite are important examples.

Ferric. Iron of a higher state of oxidation. Hematite (Fe_2O_3) is a ferric iron mineral.

Ferromagnesian. Containing iron and magnesium. Olivine, $(Fe,Mg)_2 SiO_4$, is a ferromagnesian mineral.

Ferrous. Iron of a lower state of oxidation. Siderite ($FeCO_3$) is a ferrous iron mineral.

Ferruginous. Containing iron, Fe.

Fibrous. A mineral habit consisting of threadlike individuals. Asbestos minerals show this habit to an outstanding degree.

Also, a splintery fracture in minerals.

Filiform. A mineral habit consisting of snarled, threadlike crystals. Millerite is a good example.

Fineness. The proportion of gold in native gold. If pure, it would be 1,000 fine. Vein gold in California averages 850 fine, the rest being silver. Placer gold is usually richer, because some of the silver has already dissolved away.

Fissility. The ability of a mineral to be split into sheets or layers. Mica is fissile.

Flexibility. The ability of a mineral to be bent and stay bent after the pressure is released. Among the minerals showing this property are chlorite, the so-called brittle micas, vermiculite, talc, and brucite.

Fluorescence. Illumination during exposure to radiation that is being absorbed. The fluorescence of minerals is usually carried on under ultraviolet light. The minerals transform this invisible radiation into visible colors, often very bright and of spectacular beauty. This is a useful way to identify certain minerals and gems, and to prospect for certain ores. Scheelite almost always fluoresces white or bluish white. The zinc mines near Ogdensburg, New Jersey, are famous for their fluorescent willemite (green) and calcite (red). Many diamonds fluoresce. Autunite and scapolite are other outstanding minerals that show this property.

Fluoride. A mineral composed of fluorine and another element or group of elements. Fluorite and cryolite are examples.

Foliated. A mineral habit consisting of leaves that can be separated readily. Chlorite is a good example.

Fool's gold. Chalcopyrite, pyrite, weathered biotite. Under certain circumstances, each can be mistaken for native gold. So also might other metallic minerals, such as pyrrhotite, millerite, pentlandite, and marcasite; but to the experienced prospector, native gold has a look all its own.

Fracture. The splitting of a mineral in random directions. Although it is not regular, as are cleavage and parting, it often assumes recognizable patterns. These include *conchoidal* (shell-like), *fibrous* or *splintery, hackly* (jagged), *even,* and *uneven* or *irregular.*

Friable. Crumbly.

Fusibility. Resistance of a mineral to melting. The standard (Von Kobell) scale of fusibility for the blowpipe is (from the easiest to the most difficult): 1 stibnite, 2 chalcopyrite, 3 almandite garnet, 4 actinolite, 5 orthoclase, 6 bronzite, 7 quartz.

Geiger counter. An electronic instrument for detecting and measuring radioactivity. The particles and rays given off by uranium, thorium, and other radioactive elements penetrate the Geiger-Mueller tube, yielding charged particles in the inert gas within the tube. These produce electrical impulses which are amplified—and indicated either by a meter, in earphones, or as a flashing light. Portable models have been widely used in exploration for uranium deposits.

Geode

Geode. A hollow, rounded nodule of rock containing an outer shell (like a melon) and an inner lining or filling of mineral matter. Bands of agate are common in geodes, as are crystals of quartz, calcite, and other minerals. Some geodes contain water, petroleum, or loose sand. The specimens found in the Mississippi Valley region are outstanding, and so are the gem-filled amethyst geodes from Uruguay.

Glistening. A luster that is bright but diffuse.

Granular. A mineral habit consisting of an aggregate of rounded grains. Magnetite, chromite, franklinite, olivine, and garnet are among the minerals having this habit.

Greasy. The luster of oil. Nepheline shows it, and also some specimens of quartz.

Group. A number of minerals of closely similar structure. An example is the *spinel group,* consisting of spinel, magnetite, franklinite, and chromite. Or the minerals may have chemical compositions or physical properties similar enough to warrant describing them together.

Habit. The crystal form, combination of forms, or imitative shapes in which a mineral occurs. A *massive* mineral has no particular shape.

Hackly. A fracture pattern in minerals, giving a jagged surface. The native metals, such as copper, show it most distinctly.

Halide. A mineral composed of a metal combined with one of the halogen elements, which are fluorine, chlorine, bromine, and iodine. Fluorite and cryolite are examples of the fluorides; halite, sylvite, cerargyrite, atacamite, and carnallite, of the chlorides; bromyrite, of the bromides, which are rare; iodyrite, of the iodides, also rare.

Halogen. A group of chemical elements including fluorine, chlorine, bromine, and iodine.

Hardness. The resistance of a mineral to being scratched. It should not be confused with the resistance of a mineral or other substance to being broken, because some minerals are scratched more or less easily than they can be shattered. Jade, which is very tough but not too hard, is an example. Hardness is expressed by the Mohs scale of 10 numbers. Minerals vary in hardness, but none so much as kyanite, which ranges from 5 to 7, according to the direction in which it is scratched. A penknife can be used to scratch minerals up through 5, and a penny those up to 3.

Hydroxide. A mineral containing oxygen and hydrogen as the hydroxyl ion (OH) rather than in the form of water. Hydroxide minerals yield water when heated. Brucite, manganite, and psilomelane are examples. Such minerals are often classed with the hydrous oxides, which contain water, and also with such water-yielding multiple oxides as diaspore and goethite.

Hydroxyl. A combination of one atom of hydrogen and one of oxygen (OH). Many minerals, such as tourmaline, serpentine, and the mica group, contain hydroxyl.

Hygroscopic. Capable of taking up moisture from the air.

Inclusion. A foreign mineral or other material enclosed within a mineral. Inclusions are often useful in determining the identification or source of a gem.

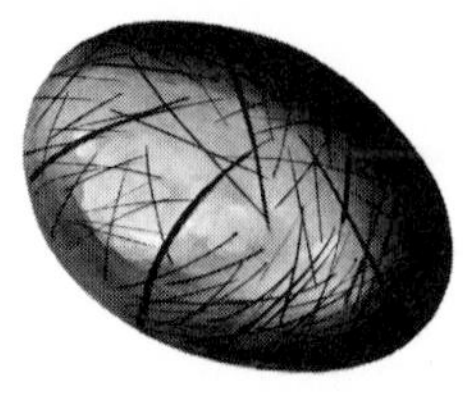

Rutile inclusions in quartz

Iodide. A mineral composed of iodine and another element or group of elements. Iodyrite is an example.

Ion. A charged atom, positive or negative. Most minerals consist of ions held together by bonding forces. The structure of most minerals is determined by the number of each kind of ion, their sizes, and their arrangement.

Iridescence. Rainbow colors on the surface of a mineral or in the interior. The cause is usually an outside, light-splitting film (often of later mineral deposition) or an internal crack or cleavage.

Jasperized wood. Petrified wood in which the filling or replacing silica has the gemmy form of jasper.

Jolly balance. A spring balance used to determine specific gravity. The specimen is weighed twice, first in air and then in water. The weight in air is divided by the loss of weight when submerged, and the figure obtained is the specific gravity. Density is expressed by the same number.

Lamellar. A mineral habit consisting of thin scales or flat plates grown upon one another.

Lime. Calcium oxide, CaO. Lime, or limy, often refers to a rock or mineral containing calcium. Thus, plagioclase is called *soda-lime feldspar*.

Luminescence. Illumination by absorbed energy. Minerals are usually made luminescent by the presence of chemical impurities or structural imperfections. According to the way they react, they may show *fluorescence, phosphorescence, thermoluminescence,* or *triboluminescence.*

Luster. The appearance of light reflected from the surface of a mineral. Minerals are divided into those having *metallic, submetallic,* and *nonmetallic* lusters. Nonmetallic lusters are further subdivided into *adamantine* (brilliant), *vitreous* (glassy), *resinous, pearly, greasy,* and *silky.*

Magnesia. Magnesium oxide, MgO. Periclase is composed solely of magnesia.

Magnetism. The ability to be attracted by a magnet. Magnetite and pyrrhotite are the two common magnetic minerals. Platinum and native iron (terrestrial or meteoritic) are also magnetic; ilmenite, franklinite, and chromite may be slightly so. A strong electromagnet will attract many other minerals. The lodestone variety of magnetite acts as a true magnet and will itself attract iron and steel.

Malleability. The ability of a mineral to hold together while being beaten. Such a mineral can be hammered or rolled. Native gold is the most malleable of all substances.

Mammillary. A mineral habit consisting of a large, rounded growth. Hematite and malachite often show this habit.

Manganate. A multiple oxide mineral containing manganese and another metal atom bonded equally to oxygen. Hausmannite, galaxite, and jacobsite are examples.

Massive. A mineral habit having no particular shape, either crystal form or imitative appearance. A mineral that cleaves into irregular pieces is said to occur as a *cleavable mass.*

Meager. A rough feel. Diatomaceous earth (fossilized deposits of diatoms' skeletons) is a good example.

Megascopic. Visible to the unaided eye. Whatever is seen in the hand specimen is *megascopic,* as opposed to *microscopic* observations.

Micaceous. A mineral habit consisting of thin sheets that can be split easily. The mica minerals show it perfectly, and it also appears in the brittle micas, the chlorite group, specular hematite, and certain other minerals.

Micromount. A tiny, mounted crystal specimen observed under magnification. A fairyland of crystalline color and form is vividly shown under the microscope. Micromounts are usually cemented onto a cork or piece of balsa wood inside a small box made of paper or plastic.

Mineralogy. The science and study of minerals. It includes their structure and physical properties, chemical composition and properties, classification, determination, and uses. Although an independent subject, it includes much of crystallography and is a useful branch of geology.

Mineraloid. A mineral substance that is not crystalline and does not have a definite chemical composition. Mineraloids form at low temperatures and pressures, usually during weathering, and occur in rounded shapes. Opal is the best example.

Mohs' scale. The standard scale of hardness in minerals. The numbers are relative only, not proportional. One of higher number will scratch any of lower number. Care should be taken not to confuse a true scratch with a mark of powder or streak. The Mohs scale is: 10 diamond, 9 corundum, 8 topaz, 7 quartz, 6 orthoclase, 5 apatite, 4 fluorite, 3 calcite, 2 gypsum, 1 talc.

Molybdate. A mineral containing the molybdate ion (MO_4). Wulfenite is the only important representative of this class.

Native element. A mineral composed of a single chemical element. About 20 elements occur as minerals, divided into *native metals, native nonmetals,* and *native semimetals.*

The metallic elements that occur as minerals include the gold group, native mercury, moschellandsbergite, the platinum group, the iron group, native tin, and native zinc. A few other minerals, especially those occurring in meteorites, fit into this class in a less definite way.

The nonmetallic elements that occur as minerals include the sulfur group and carbon group.

The semimetallic elements that occur as minerals include the arsenic group and tellurium group.

Nitrate. A mineral containing the nitrate ion (NO_3). Nitrates are easily soluble in water. Soda niter and niter are the most abundant minerals in this class.

Nugget. A lump of native metal. Gold, silver, copper, and platinum are the minerals that usually occur as nuggets. Copper nuggets from Michigan (Keweenaw Peninsula) have weighed many tons. The largest gold nugget known is the Welcome Stranger, from Ballarat, Victoria (Australia), weighing 2,280 ounces.

Ocherous. A mineral habit consisting of earthy, powdery material. Limonite is a good example.

Oölitic. A mineral habit consisting of small, rounded grains the size of fish eggs. Oölites grow around a tiny nucleus. Calcite and aragonite occur in this habit.

Opalescence. A milky or pearly reflection inside a mineral. Moonstone shows it typically.

Opalized wood. Petrified wood in which the filling or replacing silica is opal. It may be either gem or ordinary opal.

Oxide. A mineral composed of oxygen combined with a more positive element or group of elements, usually consisting of one or more metals. The oxides may be divided into *simple oxides, multiple oxides,* and other classes. Those containing hydroxyl (OH) or water (H_2O) are also known as *hydroxides.* Important oxides include: cuprite, water, zincite, magnetite, franklinite, corundum, hematite, ilmenite, braunite, rutile, pyrolusite, cassiterite, uraninite, spinel, magnetite, chromite, hausmannite, chrysoberyl, columbite, tantalite. Although quartz is chemically an oxide (SiO_2), its structure is like that of the silicates, and so it is now classed with them.

Oxysulfide. A mineral containing an element or group of elements in which part of the sulfur is replaced by oxygen. Kermesite (the alteration product of stibnite) is an example.

Paramorph. A mineral that has been derived from another one without a change in chemical composition, as when aragonite becomes calcite (both are $CaCO_3$).

Parting in pyroxene

Parting. The splitting of a mineral in directions of weakness. Unlike cleavage, it occurs only in certain specimens and even then to a limited extent. It is usually due to twinning or pressure. Pyroxene, magnetite, and corundum are among the best examples of the minerals that show parting.

Pearly. The luster of pearl. The cleavage surface of most minerals has this iridescent appearance.

Percussion figure. A pattern of radiating lines formed in a mineral by a sharp blow. Phlogopite mica shows this effect.

Petrified wood. Wood turned to mineral by filling and replacement, usually by silica. Wood may also be petrified by calcification, pyritization, and in other ways. Gem-quality wood of this sort is termed *agatized, jasperized,* and *opalized.* Few specimens equal the colorful ones in and near Petrified Forest National Park, Arizona. The known standing petrified tree stumps of largest diameter are those in Colorado (Florissant).

Phosphate. A mineral containing the phosphate ion (PO_4). Of the many phosphate minerals, the most important include xenotime, monazite, vivianite, variscite, amblygonite, apatite, pyromorphite, lazulite, turquoise, wavellite, and autunite.

Phosphorescence. Illumination after exposure to absorbed radiation. Some of the green willemite from Ogdensburg, New Jersey, phosphoresces vividly for a long time after being subjected to ultraviolet light.

Piezoelectricity. Electricity or polarity due to pressure. Piezoelectric minerals, such as quartz, lack a center of symmetry. They develop surface electric charges when rapidly oscillated, and they vibrate at controlled frequencies when an alternating current is applied. Such minerals (and artificial substitutes) are used in radio and electronic devices.

Pisolitic. A mineral habit consisting of rounded grains the size of peas. Bauxite is the typical example, and calcite and aragonite are also characterstic.

Play of color. A mixed-color effect produced by structural features in a gem. Opal is the best example.

Pleochroism. Showing different colors when viewed in different directions. It includes *dichroism* but refers mainly to the effect as observed in thin-sections of rocks under the petrographic microscope. Pleochroism is an important property in the study and identification of rock-forming minerals.

Plumose. A mineral habit consisting of fine, overlapping scales. Specular hematite sometimes shows this habit very well.

Polymorph. A mineral that has the same chemical composition as another mineral. Diamond and graphite, both native carbon, are polymorphs. So are pyrite and marcasite (FeS_2), and calcite and aragonite ($CaCO_3$). Polymorphism exists because the same atoms can be arranged in different crystalline patterns.

Potash. Potassium oxide, K_2O. The word is often used to refer to the potassium content of a rock or mineral. Orthoclase and microcline are called potash feldspar.

Precious metal. Gold, silver, platinum. Certain industrial metals may, however, be more costly than these.

Prismatic. A mineral habit consisting of an elongated crystal. Beryl, tourmaline, and corundum are among the prismatic minerals.

Property. Any characteristic or quality. Minerals and rocks are identified by their optical and other physical properties, and also by their chemical properties. Some properties, such as hardness, are directional (vectorial), and so also are the colors of most crystals.

Pseudomorph. A mineral that has the shape or crystal form of another mineral, from which it has chemically altered. It may be derived from another substance, as is petrified wood, thus forming a *cast.* Cubes of pyrite commonly change to so-called limonite (probably mostly goethite) and are described as "pseudomorphs of limonite after pyrite."

Pyramidal. A mineral habit consisting of a crystal pointed at the ends. Quartz and topaz are typically pyramidal.

Radiated. A mineral habit consisting of separate crystals that diverge from a common center. Clusters of tourmaline may show this pattern in a striking fashion. Wavellite and pectolite grow in compact, globular masses of radiating fibers.

Radioactivity. The spontaneous and uniform breakdown of atomic nuclei. The result is to emit radiation (rays and particles), producing heat, and changing the atoms to lighter ones. Uranium and thorium are the main radioactive elements, and all minerals containing them are radioactive. The end product, after radioactivity has ceased, is lead. Radioactivity can be detected with a Geiger counter or scintillation counter, by certain fluorescent effects, and on photographic film.

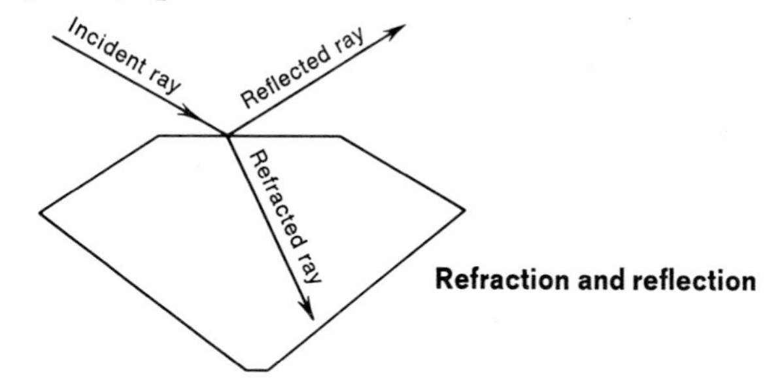

Refraction and reflection

Refraction. Bending of light. Light rays are bent or refracted when passing from one medium to another, as when entering and leaving a gem stone. The degree of bending is expressed as the *index of refraction,* which can be measured on a refractometer or determined by immersing the gem in appropriate liquids of known refraction. Gems and minerals can be identified in this way. The brilliance of diamond is largely due to its high index of refraction (2.42). Each wave length or color has a different index of refraction; thus, refracted white light may be dispersed into its constituent colors, hence the diamond's "fire." *Double refraction* is the separation of a ray of light into two rays, each of which has its own index of refraction.

Refractory. Heat resistant. A number of such minerals and rocks, such as chromite and quartzite, are used industrially.

Reniform. A mineral habit consisting of a rounded growth about the size and shape of a kidney. Hematite often shows this habit so well that it is called *kidney ore.*

Resinous. The luster of resin. Sphalerite is the typical example of a mineral that shows it.

Reticulated. A mineral habit consisting of a network of crystals. Cerussite forms such a lattice of brilliant, white crystals crossing at 60-degree angles.

Rockhound. A popular name for the mineral hobbyist of any kind. This term originated in the central United States, where it was applied to the pioneer oil-field geologists.

Scaly. A mineral habit consisting of flakes. Mica is an example.

Schiller. A bronzy, iridescent luster. Certain varieties of feldspar show this effect.

Scintillation counter. An electronic instrument for detecting and measuring radioactivity. An extremely sensitive phosphor crystal changes entering radiation into tiny flashes of light, which produce electrical impulses in a photomultiplier tube. These are amplified and indicated on a meter.

Sectility. The ability of a mineral to hold together while being cut with a knife. The selenite variety of gypsum is a good example of a sectile mineral. Certain silver minerals, especially argentite and cerargyrite, are also sectile.

Selenide. A mineral composed of selenium combined with a more metallic element or group of elements. Naumannite, umangite, clausthalite, and tiemannite are among the few selenide minerals. Selenide compounds, especially in some soils in the western

United States, may be sufficiently absorbed by certain plants as to poison grazing livestock.

Series. A sequence of minerals that shows a continuous variation in physical properties with a changing chemical composition. Thus, tetrahedrite—as it grades into tennantite—has a continuously lower specific gravity as the antimony is substituted for by the arsenic.

Silica. Silicon dioxide, SiO_2. Quartz is the most familiar form of silica.

Silicate. A mineral consisting of tetrahedral units having the composition SiO_4. These units may be independent or double, or joined to form three-dimensional networks, sheet structures, chain structures, or ring structures. These arrangements determine the fundamental properties of the silicates and their classification. Silicates are the most important minerals, making up about 25 percent of known species, almost 40 percent of the common ones, and over 90 percent of the earth's crust. Although quartz is an oxide in chemical composition, it has the silicate structure of the network type.

Siliceous. Containing silica, SiO_2.

Silky. The luster of silk. Fibrous minerals show it well; these include malachite, satin spar (gypsum), and the asbestos minerals.

Soda. Sodium oxide, Na_2O. The word is often used to refer to the sodium content of a rock or mineral. Thus, plagioclase is referred to as *soda-lime feldspar* because of its composition, which ranges from $NaAlSi_3O_8$ (albite) to $CaAl_2Si_2O_8$ (anorthite).

Species. An individual mineral of definite chemical composition. Or a major member of a series, as tetrahedrite in the tetrahedrite-tennantite series.

Specific gravity. The weight of a mineral or rock (or other substance) compared to that of an equal volume of water. Water has a specific gravity of 1. Thus, the average specific gravity of metallic minerals is about 5; of nonmetallic minerals, 2.60 to 2.75. Ice, which floats on water, has a specific gravity of 0.92. Members of the platinum group are the heaviest minerals, with specific gravities as high as 22.84. Density is expressed in the same numbers. These values are determined on a Jolly balance, beam balance, or Berman balance, in a pycnometer, or in heavy liquids.

Specimen. A rock or mineral sample. A *hand specimen* ranges up to about 3 by 4 inches; a *museum specimen* is still larger; a *thumbnail specimen* is up to about 1 inch square; a *micromount* is of microscopic size.

Specular. Having a mirrorlike luster. Specular hematite (specularite) is a typical example.

Splendent. A shining luster.

Stellated. A mineral habit consisting of separate individuals forming starlike groups.

Stone. A general name for rocky material. Properly identified, it will be a mineral or a rock. The word is useful for describing rocks and minerals used commercially as building stones, gem stones, grinding stones, and in numerous other ways.

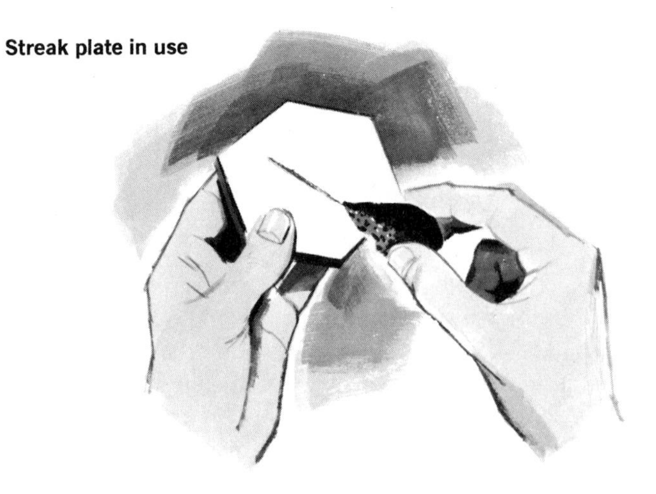

Streak plate in use

Streak. The color of a mineral when powdered. It is usually white or a lighter hue than the solid mineral. A piece of unglazed porcelain, called a streak plate, is used to rub the mineral against. Among the minerals for which streak is most useful in identification are hematite, limonite, chromite, franklinite, manganite, and psilomelane.

Streak plate. A piece of unglazed porcelain used to determine the streak of minerals.

Submetallic. A luster not quite metallic. Chromite is often submetallic.

Sulfantimonide. A mineral composed of sulfur and antimony. Ullmannite and gudmanite are examples of this chemical class of mineral.

Sulfarsenide. A mineral composed of sulfur and arsenic. Cobaltite and arsenopyrite are examples of this chemical class of mineral.

Sulfate. A mineral containing the sulfate ion (SO_4). Among the more common sulfate minerals are barite, celestite, anglesite, anhydrite, glauberite, langbeinite, polyhalite, gypsum, chalcanthite, melanterite, epsomite, brochantite, antlerite, alunite, and jarosite.

Sulfide. A mineral composed of sulfur combined with a metal or semimetal. If both are present, the semimetal replaces part or all of the sulfur in the crystal structure. The sulfides are classified into 10 types, which are subdivided into numerous groups and a few independent species. Included with the sulfides in the Dana system are the selenides and tellurides, the arsenides and antimonides, and a few oxysulfides. Among the many important and interesting sulfides are argentite, chalcocite, bornite, galena, sphalerite, chalcopyrite, pyrrhotite, covellite, cinnabar, realgar, orpiment, stibnite, pyrite, marcasite, and molybdenite. Cobaltite and arsenopyrite are arsenide-sulfides or sulfarsenides. Ullmannite and gudmanite are antimony-sulfides or sulfantimonides.

Sulfosalt. A mineral composed of a double sulfide, in which a semimetal takes the place of a metal. Thus, enargite (Cu_3AsS_4) may be regarded as a sulfide of copper (3CuS) and also as a sulfide of arsenic (AsS). Other important sulfosalts include polybasite, pearceite, stephanite, pyrargyrite, proustite, tetrahedrite, tennantite, bournonite, and jamesonite.

Tabular. A mineral habit consisting of flat plates (thicker than lamellar) grown upon one another.

Tantalate. A multiple oxide mineral containing tantalum and another metal atom bonded equally to oxygen. Tantalite, formanite, and microlite are examples.

Tarnish. A coating that conceals the true color of a mineral. Bornite shows it so typically (purple-blue to black) that the fresh color (bronzy) is seldom seen. Native silver, native copper, and other copper minerals, including chalcocite and chalcopyrite, tend to tarnish readily.

Telluride. A mineral composed of tellurium combined with a more metallic element or group of elements. The gold tellurides —such minerals as calaverite, sylvanite, krennerite, and petzite— are the only naturally occurring gold compounds. Western Australia, Rumania, Colorado, California, Ontario, and Quebec are the main sources. Silver, copper, lead, nickel, thallium, mercury, and bismuth also form telluride minerals.

Tenacity. The resistance of a mineral to being broken. Most minerals are brittle, but some are elastic, flexible, malleable, ductile, or sectile.

Thermoluminescence. Illumination caused by heating; generally, the light is given off between temperatures of 50° and 100° C. Among the minerals that sometimes show this property are fluorite, calcite, apatite, lepidolite, feldspar, and scapolite.

Thumbnail. A small mineral or rock specimen. This size of specimen has become popular with collectors because of its convenience when storing and exhibiting, which is always done using a number of such specimens at a time.

Titanate. A multiple oxide mineral containing titanium and another metal atom bonded equally to oxygen. Perovskite and pseudobrookite are examples.

Toughness. Resistance to breakage by cleaving, parting, or fracturing. Jade is a mineral that is very tough (anvils have been made of it) though not exceptionally hard.

Triboluminescence. Illumination caused by friction. Sphalerite, fluorite, and lepidolite sometimes glow in the dark when they are crushed or scratched. Most minerals showing this are nonmetallic and have good cleavage.

Tungstate. A mineral containing the tungstate ion (WO_4). The commercially important examples of this small class include huebnerite, wolframite, ferberite, and scheelite.

Type. A subdivision of a mineral class, based on a decreasing proportion of positive to negative ions or electrostatically charged atoms. For example, the spinel group has the type formula AB_2X_4, represented by $MgAl_2O_4$ for spinel, $FeFe_2O_4$ for magnetite, etc. Type is the same as *family*. Mineral types may be subdivided into *groups*.

Uranate. A mineral containing uranium and oxygen. Uraninite, the chief uranium mineral, is the best known example, but it and the others are now classed as oxides.

Vanadate. A mineral containing the vanadate ion (VO_4). The commercially important examples of this class include descloizite and vanadinite. Carnotite and tyuyamunite, of similar composition, are also classed as vanadium oxysalts because their atomic structure is less certain.

Variegated. Showing different colors or patterns.

Variety. A mineral deviating from the chemical composition of a species. Thus, freibergite is a silver-bearing variety of tetrahedrite. Or the mineral may show a physical difference; thus, ruby is red corundum. Or it may have a distinctive origin, as travertine is a spring deposit (hot or cold) of calcite.

Vitreous. The luster of glass. Quartz and many other minerals show it.

X-ray diffraction. A method of identifying minerals and other crystals by means of the distinctive pattern produced by the bending of X-rays as they pass through the layers of atoms in the substance. It was begun in 1912 and has become the most certain method of identification. There are three principal techniques: the powder method (using powdered material), the rotation method (using a moving crystal), and the Laue method (using a stationary crystal).

Zirconate. A multiple oxide mineral containing zirconium and another metal atom bonded equally to oxygen. Zirkelite is an example.

CRYSTALS

A crystal is a solid body composed of a chemical element or compound and showing flat surfaces or faces that are symmetrically arranged. These faces express a regular and distinctive atomic structure characteristic of the element or compound. Crystals are formed from solution, fusion, or vapor. Minerals develop as crystals when conditions are favorable, usually when open space is provided in the rocks. Crystals are grouped into 6 **systems,** 14 **space lattices,** 32 **classes** or **point groups,** and 230 **space groups.**

Acute. Sharply pointed.

Albite twin. A triclinic twin law involving a twin plane on the side pinacoid. Plagioclase feldspar almost always crystallizes as this kind of twin, which produces striations on the best cleavage surface.

Axis. An imaginary line passing through the center of a crystal. Hexagonal crystals have four axes; all others have three.

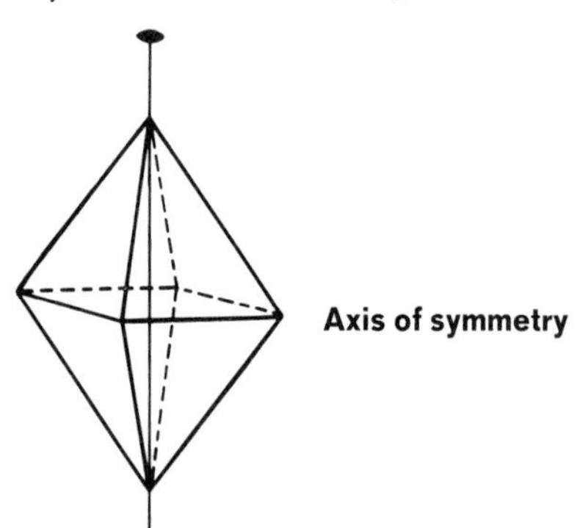

Axis of symmetry. An imaginary line through a crystal. The crystal may be presumed to rotate around this axis, producing a repetition of faces. Thus, in threefold symmetry, the same face is seen three times in one complete rotation of the crystal.

Basal. Parallel to the top and bottom of a crystal.

Baveno twin. A monoclinic and triclinic twin law involving a twin plane on the first-order prism or side pinacoid. Orthoclase and microcline feldspar often crystallize as this kind of twin, which yields a nearly square form.

Carlsbad twin. A monoclinic and triclinic twin law involving a twin plane on the side pinacoid. Orthoclase and microcline feldspar often crystallize as this kind of twin.

Center of symmetry. A point in the center of a crystal. Any straight line through the center connects exactly similar points at opposite ends. Crystals that lack a center of symmetry can show piezoelectricity, transforming oscillating pressure into alternating current, and vice versa.

Class. One of the 32 ways in which crystals are divided on the basis of their planes, axes, and center of symmetry. It is the same as point group.

Closed form. A crystal form that entirely encloses space by itself. Thus (unlike an open form), a cube does not need any other form with it to make a complete crystal.

Composition plane. The surface joining the parts of a contact twin. It is usually the same as the twin plane.

Compound twin. A twin crystal having the individuals united according to more than one twin law.

Contact twin. A twin crystal whose parts are joined along a surface called the composition plane. If the twin consists of three or more parts twinned according to the same law, it is a *repeated,* or *multiple, twin.* If the composition planes are parallel, it is a *polysynthetic twin;* if not, it is a *cyclic twin.*

Cruciform. Cross shaped. The twin crystals of staurolite are cruciform.

Cryptocrystalline. Very finely crystalline. The individual grains or crystals cannot be distinguished without X-rays. Chalcedony is cryptocrystalline quartz.

Crystalline. Having an orderly atomic structure, whether or not the geometric form of a crystal is present. All solid minerals are crystalline except opal and other mineraloids.

Crystallography. The science and study of crystals and crystalline substances. It is an independent subject, but much of it is necessary to a knowledge and appreciation of mineralogy.

Cube. An isometric crystal form consisting of six faces at right angles to each other. Each face cuts one axis and is parallel to the other two axes. When complete, each face is a square. Pyrite, galena, halite, and fluorite are among the minerals commonly showing this form.

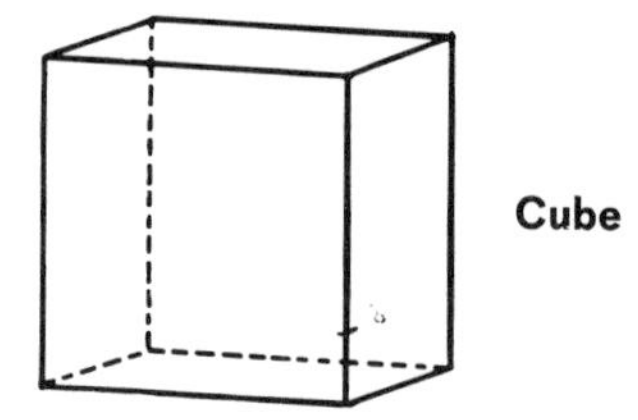

Cyclic twin. A repeated, or multiple, twin whose composition planes are not parallel. Handsome, spokelike crystals of chrysoberyl show this kind of twinning, as do also the common, disklike twins of aragonite known as Indian dollars.

Deltohedron. An isometric crystal form consisting of 12 faces. Each face cuts two axes equally and the third axis at a different length. When complete, each face is a trapezium (four sides, no two parallel). This form is also called a deltoid dodecahedron.

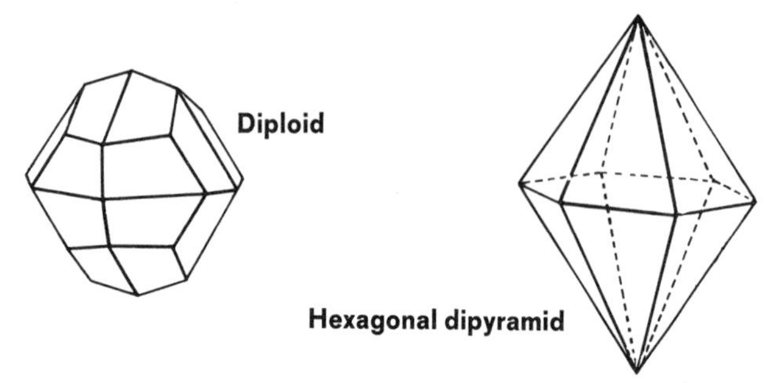

Diploid. An isometric crystal form consisting of 24 faces. Each face cuts all three axes at different lengths. When complete, each face is a trapezium (four sides, no two parallel). Pyrite often shows this form in combination with others.

Dipyramid. A crystal form consisting of a number of faces that cut the vertical axis. There may be 6, 8, 12, 16, or 24 faces. According to their symmetry, dipyramids are named *hexagonal, dihexagonal, trigonal, ditrigonal, tetragonal, ditetragonal,* or *rhombic.* They occur only on crystals in the tetragonal, orthorhombic, and hexagonal systems.

Disphenoid. A tetragonal or orthorhombic crystal form consisting of four faces, two upper ones alternating with two lower ones. Each face cuts all three axes. When complete, each face is an isosceles triangle (two sides equal). Chalcopyrite commonly shows this form.

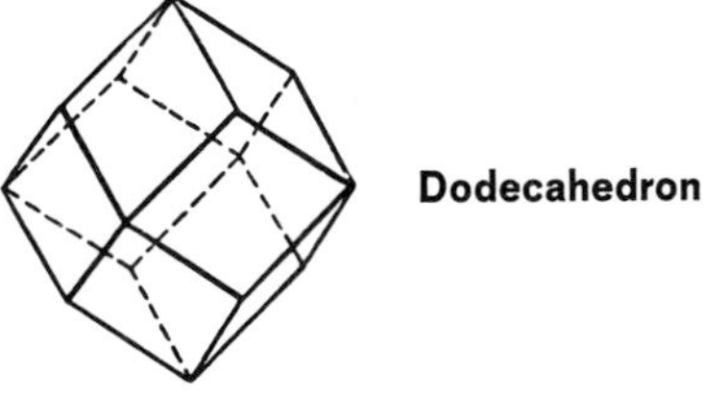

Dodecahedron

Dodecahedron. An isometric crystal form consisting of 12 faces. Each face cuts two axes equally and is parallel to the third axis. When complete, each face is a rhomb (four sides, with opposite sides parallel). Sphalerite and garnet are among the minerals commonly showing this form. A deltoid dodecahedron is a rare, isometric crystal form of different symmetry.

Dome. An orthorhombic or monoclinic crystal form consisting of two nonparallel faces that are symmetrical with respect to a plane of symmetry. It is almost identical with a sphenoid.

Elbow twin. A tetragonal twin law involving a twin plane on the second-order dipyramid. Rutile often crystallizes as this kind of twin.

Face. The flat surface on a crystal.

Form. The association of crystal faces, all of which have the same relationship to the aspects of crystal symmetry. Thus, all six faces of a cube belong to one form. Form, therefore, does not mean simply shape. There are 48 different crystal forms. A *general form* cuts all the axes of a crystal at different lengths. A *closed form* entirely encloses space by itself; an *open form* requires other forms to be present to make a complete crystal.

General form. A crystal form that cuts all the axes of a crystal at different lengths. It customarily gives its name to the crystal class in which it occurs, as the *gyroid* does to the *gyroidal* class of the isometric system.

Hemimorphic. Having different forms at opposite ends of a crystal axis. Tourmaline is an example.

Hexagonal. A crystal system having four axes, the three horizontal ones being 120 degrees apart and of the same length, the vertical one being different. The minerals that form in good hexagonal crystals include quartz, beryl, tourmaline, calcite, dolomite, rhodochrosite, siderite, corundum, vanadinite, pyromorphite, mimetite, chabazite, dioptase, apatite, hematite, nepheline, and phenacite. This system may be divided into the *hexagonal division* (sixfold symmetry) and the *rhombohedral division* (threefold symmetry).

Hexagonal crystal axes

Hexoctahedron

Hexoctahedron. An isometric crystal form consisting of 48 faces. Each face cuts the three axes at different lengths. When complete, each face is a scalene triangle (all three sides different). Garnet often shows this form in combination with a dodecahedron.

Hextetrahedron. An isometric crystal form consisting of 24 faces. Each face cuts all three axes at different lengths. When complete, each face is a scalene triangle (all three sides different).

Indices. The numbers that express the relationship between a crystal face and the axes of the crystal. Miller indices are those most often used.

Intergrowth. The penetration of two crystals. These are usually presumed to have grown at the same time.

Iron cross twin. An isometric twin law involving penetration pyritohedrons. Pyrite often crystallizes as this kind of twin.

Isometric. A crystal system having three axes of equal lengths, intersecting in the center at right angles. The minerals that form

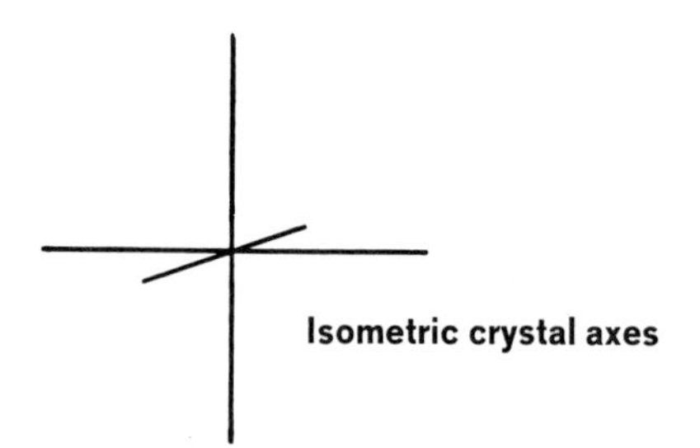
Isometric crystal axes

in good isometric crystals include diamond, garnet, halite, sylvite, fluorite, analcime, leucite, boracite, spinel, gahnite, sphalerite, cuprite, pyrite, franklinite, magnetite, and galena.

Isomorphism. The substitution of one atom for another in the crystal structure of a mineral. This is done without changing the crystal form. It involves solid solution, an example of which is the substitution of iron for magnesium in garnet, which in this way can be either almandite, $Fe_3Al_2(SiO_4)_3$, or pyrope, $Mg_3Al_2(SiO_4)_3$, or rhodolite (chemically in between).

Manebach twin. A monoclinic and triclinic twin law involving a twin plane on the basal pinacoid. Orthoclase and microcline sometimes crystallize as this kind of twin.

Microcrystalline. Finely crystalline. The individual grains or crystals are not visible without a microscope.

Miller indices. A set of three or four whole numbers that express the relationship between a crystal face and the axes of the crystal. They are obtained by measuring the parameters, inverting them, and clearing fractions. Examples are (101)—read *one-oh-one*—for a cube face; $(21\bar{3}1)$—read *two-one-minus three-one*—for a scalenohedron face common on calcite.

Modification. A crystal face that cuts other faces.

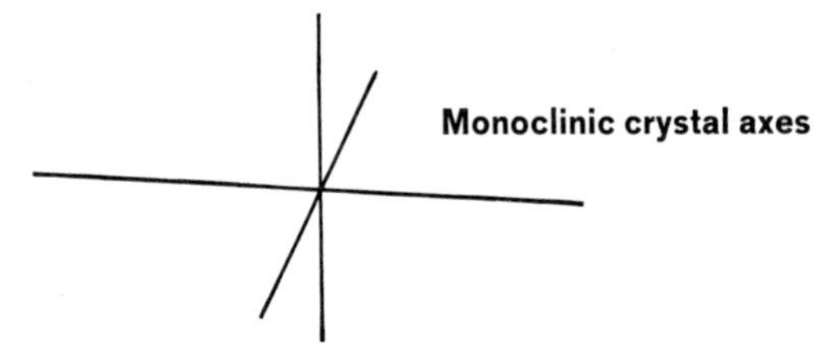

Monoclinic crystal axes

Monoclinic. A crystal system having three angles of different lengths, one being at right angles to the plane of the other two, which intersect in the center at oblique angles. The minerals that form in good monoclinic crystals include borax, heulandite, gypsum, orthoclase, datolite, amphibole, pyroxene, monazite, and staurolite.

Morphology. The external geometry of crystals.

Multiple twin. A twin crystal consisting of three or more parts twinned according to the same law. It is the same as repeated twin.

Octahedron. An isometric crystal form consisting of eight faces that meet in a double pyramid. Each face cuts all three axes at the same length. When complete, each face is an equilateral triangle (all three sides equal). Diamond, spinel, magnetite, and franklinite are among the minerals commonly showing this form.

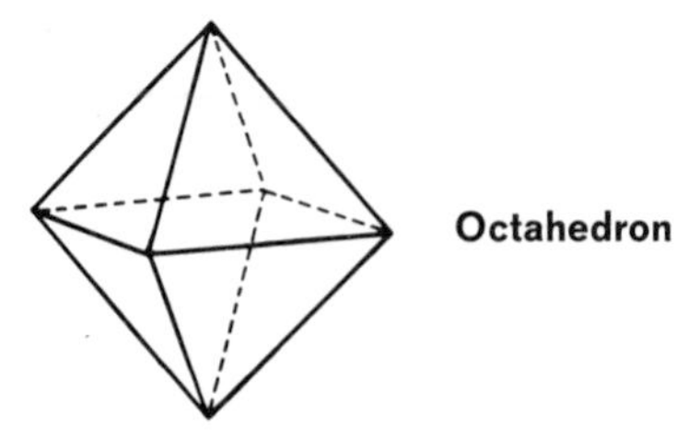
Octahedron

Open form. A crystal form that does not enclose space by itself. Unlike a closed form, it needs one or more other forms with it to make a complete crystal. Thus, a pinacoid is an open form because it is incomplete in itself.

Order. Generally speaking, order refers to the orientation of a crystal's axes and faces. For example: in a first-order prism in the hexagonal system, the three horizontal axes intercept the "corners" of the prism; in a second-order prism, the three axes intercept the faces of the prism.

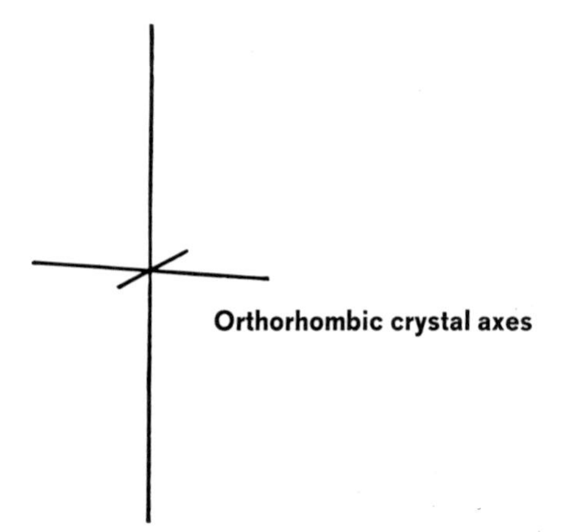
Orthorhombic crystal axes

Orthorhombic. A crystal system having three axes of different lengths, intersecting in the center at right angles. The minerals that form in good orthorhombic crystals include sulfur, aragonite, danburite, andalusite, topaz, chrysoberyl, staurolite, brookite, barite, anglesite, stibnite, marcasite, arsenopyrite, enargite, manganite, columbite, and tantalite.

Parameters. The relative distances from the center of a crystal to the intersections of a crystal face with the crystal axes. The axes are taken in a regular order. Parameters are usually converted into Miller indices by inverting them and clearing fractions.

Pedion. A crystal form consisting of only one face. It is like half of a pinacoid. Zincite commonly shows this form.

Penetration twin. A twin crystal whose parts appear to penetrate one another. The iron-cross twin of pyrite is a good example.

Pericline twin. A triclinic twin law involving a crystal axis as the twin axis. Plagioclase feldspar often crystallizes as this kind of twin, which produces striations on the second-best cleavage surface.

Phantom. Growth lines visible inside a crystal. Quartz and other minerals often show phantoms.

Pinacoid. A crystal form consisting of two parallel faces. According to their symmetry, pinacoids are named *basal, front, side, first order, second order, third order,* or *fourth order.* They occur on crystals in all systems except the isometric.

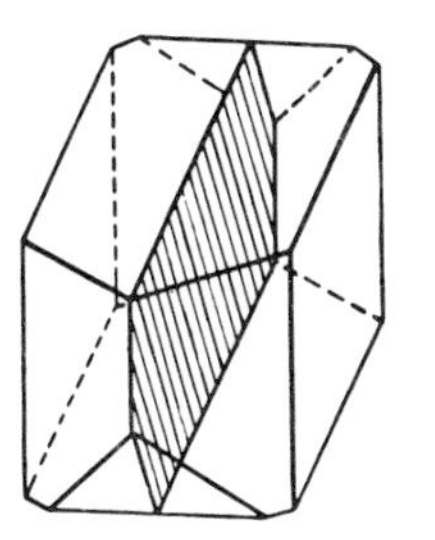

Plane of symmetry

Plane of symmetry. An imaginary, mirrorlike surface in a crystal which appears to reflect the opposite side, producing a repetition of each corner, edge, and face.

Point group. One of the 32 ways in which crystals are divided on the basis of their planes, axes, and center of symmetry. It is the same as class.

Polar. Having different crystal faces or properties at opposite ends of a crystal axis. Tourmaline is a good example.

Polysynthetic twin. A repeated, or multiple, twin whose composition planes are parallel. This kind of twinning yields the familiar striations on plagioclase feldspar.

Prism. A crystal form consisting of a number of faces parallel to the same axis. There may be 3, 4, 6, 8, or 12 faces. According to their symmetry, prisms are named *hexagonal, dihexagonal, trigonal, ditrigonal, tetragonal, ditetragonal,* or *rhombic.* They occur on crystals in all systems except the isometric and triclinic.

Pyramid. A crystal form consisting of a number of nonparallel faces that meet at a point. There may be 3, 4, 6, 8, or 12 faces. According to their symmetry, pyramids are named *hexagonal, dihexagonal, trigonal, ditrigonal, tetragonal, ditetragonal,* or *rhombic.* When they occur, as is usual, at both ends of a crystal, pyramidal forms are named *dipyramids.*

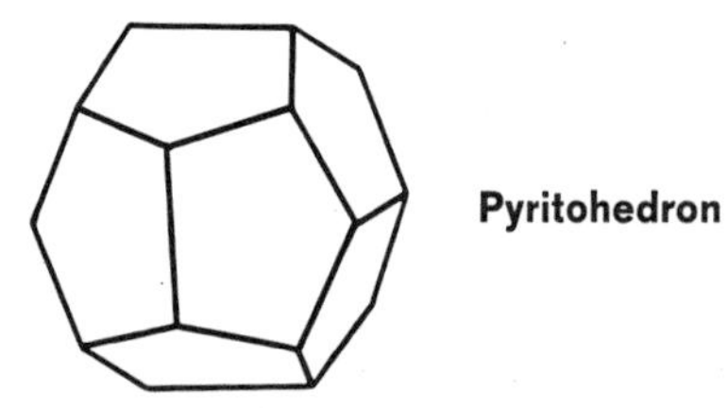

Pyritohedron. An isometric crystal form consisting of 12 faces. Each face cuts two axes at different lengths and is parallel to the third axis. When complete, each face has five sides. Pyrite and cobaltite commonly show this form.

Repeated twin. A twin crystal consisting of three or more parts twinned according to the same law. It is the same as multiple twin.

Rhombohedral division. The division of the hexagonal crystal system that includes those crystals having threefold symmetry.

Rhombohedron. A hexagonal crystal form consisting of six faces and resembling a deformed cube. Each face cuts three axes and is parallel to the fourth. When complete, each face is a rhomb (four sides, with opposite sides parallel). Quartz, calcite, dolomite, and siderite are among the minerals commonly showing this form.

Scalenohedron. A hexagonal or tetragonal crystal form consisting of 12 or 8 faces occurring in symmetrical pairs. When complete, each face is a scalene triangle (all three sides different). Calcite commonly shows this form.

Skeletal. A crystal growth having an open or cavernous appearance. Sylvanite often shows this habit.

Space group. One of the 230 ways in which crystals are divided on the basis of their internal symmetry.

Space lattice. The structure, or three-dimensional network arrangement, of atoms or groups of atoms in a crystal. The unit of structure, or *unit cell,* is the smallest portion of the crystal possessing its characteristic properties.

Sphenoid. A monoclinic crystal form consisting of two non-parallel faces that are symmetrical with respect to an axis of symmetry. It is almost identical with a dome.

Spinel twin. An isometric twin law involving contact octahedrons. Spinel and diamond often crystallize as this kind of twin.

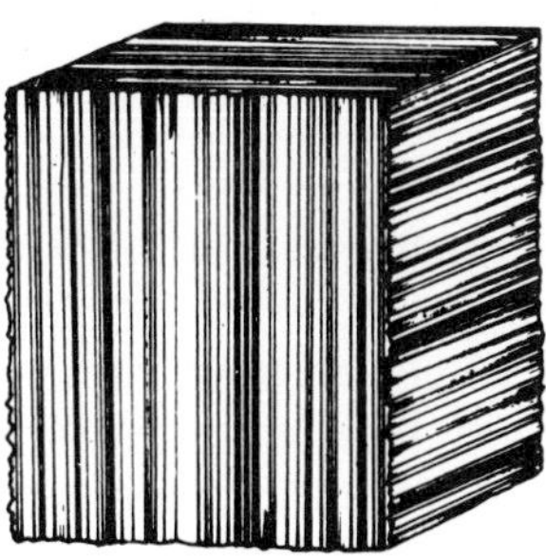

Striations on pyrite

Striations. Fine, parallel lines on a crystal. They are due either to twinning or to alternation of different crystal forms. Plagioclase feldspar can be told from orthoclase or microcline by its striations. Quartz has horizontal striations, while tourmaline has vertical ones.

Swallowtail twin. A monoclinic twin law involving a twin plane on the front pinacoid. Gypsum often crystallizes as this kind of twin.

Symmetry. The repetition of pattern in crystallography, usually in relation to planes, axes, and a center of symmetry. These are known as the *elements of symmetry.*

System. One of the six ways in which crystals are divided on the basis of the crystal axes.

Termination. The crystal faces at an end of a crystal axis.

Tetartoid. An isometric crystal form consisting of 12 faces. Each face cuts all three axes at different lengths. When complete, each face is five sided.

Tetragonal. A crystal system having three axes intersecting in the center at right angles, the two horizontal ones being the same length, the vertical one being different. The minerals that form in good tetragonal crystals include zircon, idocrase, scapolite, octahedrite, rutile, scheelite, wulfenite, cassiterite, chalcopyrite, anatase, and apophyllite.

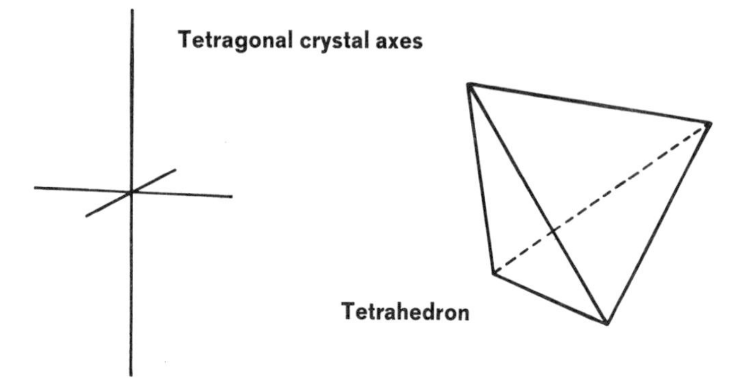

Tetragonal crystal axes

Tetrahedron

Tetrahedron. An isometric crystal form consisting of four faces. Each face cuts all three axes equally. When complete, each face is an equilateral triangle (all three sides equal). Tetrahedrite and tennantite commonly show this form.

Tetrahexahedron. An isometric crystal form consisting of 24 faces. Each face cuts two axes at different lengths and is parallel to the third axis. When complete, each face is an isosceles triangle (two sides equal). Native copper commonly shows this form.

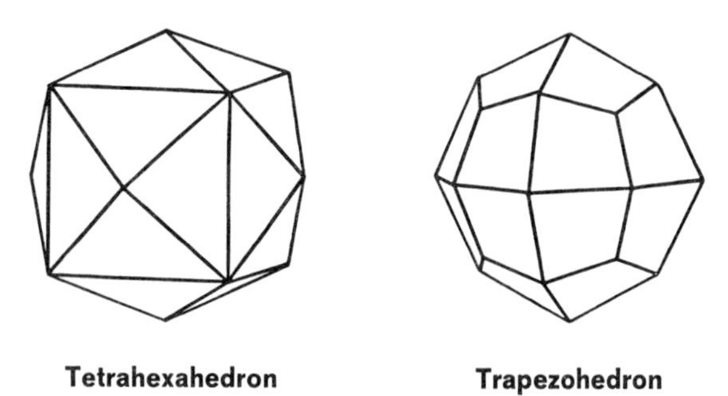

Tetrahexahedron

Trapezohedron

Trapezohedron. An isometric crystal form consisting of 24 faces. Each face cuts three axes, two of them at the same length. When complete, each face is a trapezium (four sides, no two parallel). Garnet, leucite, and analcime are among the minerals commonly showing this form.

Also, a crystal form in the hexagonal and tetragonal systems, consisting of offset pairs of faces numbering 6, 8, or 12. The trigonal trapezohedron is a six-faced form important on quartz; it helps, when its position is noticeable, to distinguish right-handed crystals from left-handed ones.

Triclinic. A crystal system having three axes of different lengths, intersecting in the center at oblique angles. The minerals that form in good triclinic crystals include microcline, rhodonite, axinite, and albite.

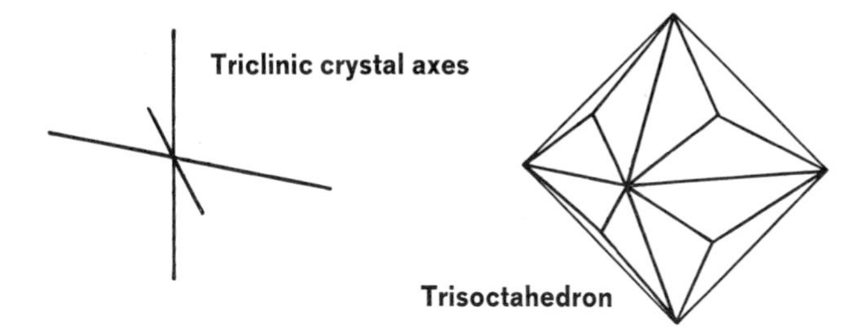

Triclinic crystal axes

Trisoctahedron

Trisoctahedron. An isometric crystal form consisting of 24 faces. Each face cuts two axes equally and the third axis at a different length. When complete, each face is an isosceles triangle (two sides equal).

Tristetrahedron. An isometric crystal form consisting of 12 faces. Each face cuts two axes equally and the third axis at a different length. When complete, each face is an isosceles triangle (two sides equal).

Twin axis. An imaginary axis common to both parts of a twin crystal, around which rotation appears to have taken place to cause a duplication of the same faces, edges, and corners. It is at right angles to any twin plane that may be involved.

Twin crystal. Two or more crystals intergrown in a definite way. The parts are related as if they were reflected across a twin

Penetration twin of fluorite

plane, or rotated around a twin axis, or symmetrical about a center of symmetry. Probably the most obvious kinds of twin crystal are the crosses of staurolite. A twin may be of *contact* or *penetration* type.

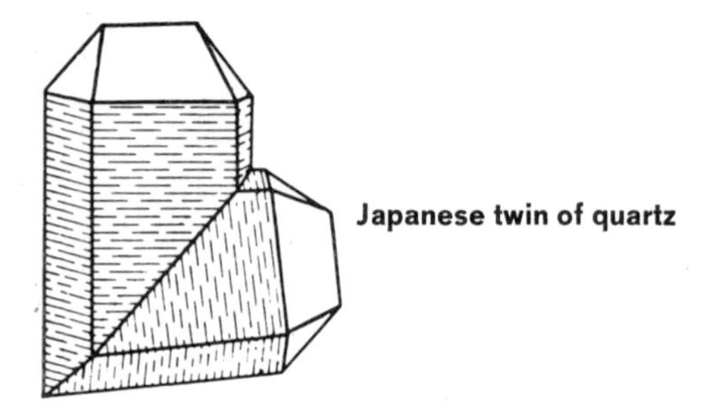
Japanese twin of quartz

Twin law. The description of the crystallography of a twin crystal. Some of the most familiar twin laws, which are given specific names, are *spinel twin* in diamond and spinel, *iron-cross twin* in pyrite, *visor twin* in cassiterite, *elbow twin* in rutile, *swallowtail twin* in gypsum, *Carlsbad twin* in orthoclase and microcline, *Manebach twin* in orthoclase and microcline, *Baveno twin* in orthoclase and microcline, *albite twin* in microcline and plagioclase, *pericline twin* in microcline and plagioclase, *Brazil twin* in quartz, *Dauphiné twin* in quartz, *Japanese twin* in quartz, and *butterfly twin* in calcite.

Twin plane. A surface common to both parts of a twin crystal, across which reflection (a mirror image) appears to have taken place. It is at right angles to any twin axis that may be involved. In a contact twin, it is also the composition plane.

Visor twin. A tetragonal twin law involving a twin plane on a second-order dipyramid. Cassiterite often crystallizes in visor twins, thus helping miners to recognize it when it is embedded in rock.

Zone. A series of crystal faces having parallel edges.

ORIGIN OF MINERAL NAMES

The familiar ending *ite,* used with most mineral names, goes back to ancient times. In the form of *ites* and *itis,* it was added by the Greeks, and later by the Romans, to ordinary words that denoted qualities, uses, constituents, or localities of minerals and rocks. Thus, "siderites" (an iron-bearing mineral now called siderite) was made from the Greek word for iron; "haematites" (now hematite) was made from the word for blood, because of the red color of the powdered mineral. All the present kinds of mineral names, except names of persons, were used in classical days.

The first mineral to be named after an individual was prehnite, which in 1783 was named by Abraham Gottlob Werner after its discoverer, Colonel von Prehn, who had brought the first specimens to Europe from the Cape of Good Hope. This method of naming a mineral has now become the most widely favored.

The ending *lite* is assumed to have come from the French suffix *lithe,* derived in turn from *lithos,* the Greek word for stone. In mineral names, however, *lite* may merely stand for *ite,* the consonant being added for euphony. Other terminations have been less extensively used, but some of the most attractive sounding mineral names have them, including tourmaline, pyroxene, euclase, epidote, and harmotome.

In 1837 James D. Dana published the first edition of his *System of Mineralogy,* in which he applied a multiple Latin name to minerals, similar to the names used in botany and zoology. In the third edition, published in 1850, Dana rejected his entire nomenclature and used single names, as we do today.

In general, the law of priority is applicable to mineral names and is superseded only when the original name is proved to have been incorrect or inadequate. In recent years, through international agreement among the English-speaking nations, certain names have been discredited in favor of others that often are less well known.

Some of the most common mineral names are so old that their origin is unknown or doubtful. Included among them are zircon, cinnabar, galena, gypsum, corundum, beryl, quartz, and tourmaline.

A variety of interesting mineral names of known origin—both ancient and modern—is given in the following list.

Mythology

Aegirite: after Aegir, Teutonic god of the sea.

Castorite and Pollucite: (because they occur together) after Castor and Pollux, the twin sons of Zeus and Leda.

Martite: after Mars, Roman god of war, because of its red streak.

Mercury: after Mercurius, Roman messenger of the gods, because of its fluid and volatile (quicksilver) nature.

Neptunite: after Neptunus, Roman god of the sea.

Tantalite: after Tantalus (condemned by Zeus to eternal thirst), because of the difficulty of dissolving it in acid.

Thorite: after Thor, Norse god of thunder.

Chemical Composition

Aenigmatite: from Greek "riddle," because of its puzzling nature.

Anhydrite: from Greek "without water," because of the absence of water, in contrast to gypsum (which is hydrous calcium sulfate).

Boracite: from English "borax"—it is magnesium borate.

Calcite: from Latin "lime"—it becomes lime (calcium oxide) when heated.

Calomel: from English "calomel"—the original name for the medical preparation having the same composition (mercury chloride).

Cerussite: from Latin "white lead"—it is a lead mineral having a white color.

Chalcopyrite: from Greek "copper" and "pyrite"—it has the composition of pyrite (iron sulfide) plus copper.

Dyscrasite: from Greek "bad mixture"—it contains antimony, an undesirable (misleading or contaminating) element to the metallurgist.
Embolite: from Greek "intermediate," because it combines both silver chloride and silver bromide, being chemically between.
Halite: from Greek "salt"—it is common salt (sodium chloride).
Magnesite: from English "magnesium"—it is magnesium carbonate.
Miargyrite: from Greek "less" and "silver"—it contains less silver than pyrargyrite or proustite.
Mirabilite: from "sal mirabile," the expression of surprise used by J. R. Glauber when he discovered the artificial salt (before it was found as a mineral).
Natrolite: from Latin "sodium"—it is sodium silicate.
Niccolite: from English "nickel"—it is nickel arsenide.
Polybasite: from Greek "much" and "base"—it contains a large amount of two metallic bases (silver, copper).
Pyrargyrite: from Greek "fire" and "silver," because of its red color and silver content.
Thaumasite: from Greek "to be surprised," because of its unexpected composition.
Triphylite: from Greek "three" and "family"—it contains three metallic bases (manganese, lithium, and iron).
Uraninite: from English "uranium"—it is a uranium mineral.
Xenotime: from Greek "vain" and "honor," because it contains yttrium, which had been mistaken for a new element.
Zincite: from English "zinc," because it is zinc oxide.

Use

Agalmatolite: from Greek "statue"—it is often carved into images in the Orient.
Alunite: from Latin "alum"—it is used for producing alum.
Amethyst: from Greek "not drunken"—it was thought to prevent intoxication when worn.
Chrysocolla: from Greek "gold" and "glue"—it was used to solder gold.
Graphite: from Greek "to write"—it is used in pencils.
Gypsum: from Greek "plaster"—it is made into plaster of paris by burning.

Muscovite: from English "Muscovy glass"—it was used in Russia for window panes.

Nephrite: from Greek "kidney stone"—it was used as a remedy for kidney diseases.

Orpiment: from Latin "gold paint"—it was used as a gold-yellow pigment.

Pyrolusite: from Greek "fire" and "to wash"—in molten glass it removes the color due to iron impurities.

Smaltite: from English "smalt," a blue pigment made with the powdered mineral.

Cleavage

Dioptase: from Greek "to see into"—its cleavage planes can be seen within the crystal.

Euclase: from Greek "to break well," because of its perfect cleavage.

Microcline: from Greek "small" and "to incline"—the angle between its two chief cleavages is almost a right angle.

Oligoclase: from Greek "little" and "to break"—it was believed to have a less perfect cleavage than albite.

Orthoclase: from Greek "right" and "to break"—its two chief cleavages are at right angles.

Petalite: from Greek "leaf," because of its leaflike cleavage.

Hardness

Diamond: from Greek "invincible"—it was believed to be indestructible.

Hypersthene: from Greek "very" and "strong"—it is harder than hornblende.

Pyrite: from Greek "fire," because it strikes fire with steel.

Specific Gravity

Barite: from Greek "heavy," because of its high specific gravity.

Electricity and Magnetism

Analcime: from Greek "weak," because of its weak electrical power when rubbed.

Magnetite: from English "magnet," because of its strong magnetic power.

Color

Albite: from Latin "white."
Anthophyllite: from Latin "clove," because of its clove-brown color.
Aquamarine: from Latin "sea water," because of its blue-green color.
Azurite: from Persian "blue."
Bronzite: from English "bronze."
Celestite: from Latin "heavenly," because of the blue color of the first specimen found.
Crocoite: from Greek "saffron," because of its orange-yellow color.
Cryolite: from Greek "ice," because it is white.
Eosphorite: from Greek "dawn bearing," because of its pink color.
Glaucophane: from Greek "greenish blue" and "to appear."
Grossularite: from Latin "gooseberry," because of its yellow color.
Helvite: from Greek "sun," because of its yellow color.
Lazulite: from Arabic "heaven," because of its blue color.
Malachite: from Greek "mallow," because of its green color.
Melaconite: from Greek "black dust," because of its powdery-black color.
Melilite: from Greek "honey," because of its honey-yellow color.
Olivine: from English "olive," because of its olive-green color.
Psilomelane: from Greek "smooth" and "black," because of its surface appearance.
Pyrope: from Greek "fiery," because of its fire-red color.

Luster

Augite: from Latin "brightness," because of its bright luster.
Cerargyrite: from Greek "horn" and "silver," because of its horn-like luster and silver content.
Collophane: from Greek "glue" and "to appear," because of its pasty luster.
Margarite: from Greek "pearl," because of its pearly luster.
Stilbite: from Greek "to glitter," because of its bright luster.

Heat Effects

Apophyllite: from Greek "detached" and "leaf," because it opens up when heated.

Asbestos: from Greek "unquenchable"—it is incombustible.

Diaspore: from Greek "to scatter"—it breaks apart when heated.

Enstatite: from Greek "opponent"—it is infusible.

Fluorite: from Latin "to flow"—it melts easily.

Pyromorphite: from Greek "fire" and "form," because the globule obtained after melting has a crystalline shape.

Scolecite: from Greek "worm"—it curls up when heated.

Spodumene: from Greek "burned to ashes"—it becomes an ash-colored mass when heated with a blowpipe.

Vermiculite: from Latin "little worm"—it exfoliates into twisted threads when heated.

Zeolite: from Greek "to boil"—it bubbles when heated with a blowpipe.

Occurrence

Emplectite: from Greek "entwined," because it is so closely associated with quartz.

Monazite: from Greek "to be solitary," because of its rarity.

Pyroxene: from Greek "fire" and "stranger"—it was erroneously believed not to occur in igneous rocks.

Realgar: from Arabic "powder of the mine," because it came from a silver mine.

Localities

Alabandite: after Alabanda, an ancient city in Turkey, its supposed locality.

Alaskaite: after Alaska mine, San Juan County, Colorado.

Andalusite: after Andalusia Province, in Spain.

Anglesite: after Anglesea, an island off Wales.

Antigorite: after Antigorio Valley, in Piedmont Province, Italy.

Aragonite: after Aragon, a former kingdom in Spain.

Atacamite: after Atacama Desert, in Chile.

Calaverite: after Calaveras County, California.

Cristobalite: after Cerro San Cristobal, in Mexico.

Cubanite: after Cuba.

Epsomite: after Epsom, England.

Ilvaite: after Ilva, the old name for the island of Elba.

Jarosite: after Barranco Jarosa, in Spain.

Labradorite: after Labrador, Canada.

Sylvanite: after Transylvania, a region in Rumania.

Thulite: after Thule, the old name for Norway.

Turquoise: after Turkey, from whence it was brought to Europe.

Vesuvianite: after Mount Vesuvius, Italy.

Persons

Aguilarite: after P. Aguilar, superintendent of the San Carlos mine in Mexico, where it was first found.

Alexandrite: after Alexander II, czar of Russia.

Andradite: after J. B. d'Andrada, who first studied it.

Biotite: after J. B. Biot, the French physicist who studied its optical properties.

Bornite: after Ignaz von Born, Austrian mineralogist, who first noticed it.

Bournonite: after Count J. L. de Bournon, who first described it.

Brochantite: after A. J. M. Brochant de Villiers, French geologist and mineralogist.

Carnotite: after Marie-Adolphe Carnot, French mining engineer and chemist.

Colemanite: after William T. Coleman, owner of the mine where it was found.

Cordierite: after P. L. Cordier, French geologist, who described it.

Descloizite: after Alfred L. O. L. Des Cloizeaux, French mineralogist, who first described it.

Dolomite: after Déodat Dolomieu, the French geologist who first examined it.

Dumortierite: after E. Dumortier, French paleontologist.

Gibbsite: after Colonel George Gibbs, a mineral collector.

Goethite: after Johann Wolfgang von Goethe, German poet and natural philosopher.

Greenockite: after Lord Greenock, later Earl Cathcart.

Hanksite: after Henry G. Hanks, state mineralogist of California, where it was found.

Hiddenite: after William E. Hidden, owner of the property where it was first found.

Humite: after Sir Abraham Hume, of London.

Millerite: after William H. Miller, English crystallographer.

Pentlandite: after J. B. Pentland, who first brought it to notice.

Scheelite: after Karl Wilhelm Scheele, the Swedish chemist who first discovered tungsten, which it contains.

Smithsonite: after James Smithson, founder of the Smithsonian Institution, who distinguished it from hemimorphite.

Stephanite: after Archduke Stephan of Austria.

Swedenborgite: after Emanuel Swedenborg, Swedish philosopher.

Torbernite: after Torbern Bergman, the Swedish chemist who first examined it.

Uvarovite: after Count S. S. Uvarov, president of the Academy of St. Petersburg.

Valentinite: after Basil Valentine, an alchemist, who discovered the properties of antimony, which it contains.

Vivianite: after J. G. Vivian, the English mineralogist who discovered it.

Wavellite: after Dr. William Wavell, English physician, who discovered it.

Willemite: after William I, king of the Netherlands.

Zoisite: after Baron Zois von Edelstein, the Austrian mineral collector who discovered it.

MINERALS AND GEMS

A mineral is a natural chemical substance of inorganic origin. It may be an element or a compound and is always homogeneous. Man-made substances may be identical with minerals, and the organic activity of plants and animals produces some of the same products, but they are not true minerals. There are from 1,600 to 2,000 species of minerals, the number depending on the way they are classified. (The Dana system of classification is used in this book.) Most minerals are crystalline and occur in crystals when conditions are favorable. They have recognizable physical properties (such as hardness and specific gravity), which are variable within definite limits, according to the chemical composition.

A gem is a mineral or other substance that is used for adornment. It is beautiful, durable, or rare, and often all of these. Some gems are rocks, some are organic substances, and some are of artificial origin. Ornamental stones are minerals and rocks that are used also for decoration but are less valuable than the recognized gems.

Acadialite. The local name in Nova Scotia for chabazite.

Acicular bismuth. An old name for aikinite, because of its needlelike crystals and metal content.

Acmite. A variety of aegirite. It occurs in long, slender crystals. Sometimes the distinction is made on the basis of the color, acmite being brown rather than green.

Actinolite. A member of the tremolite-actinolite series of minerals in the amphibole family. It is a green and vitreous hydrous silicate, $Ca_2(Mg,Fe)_5(Si_8O_{22})(OH)_2$. With a decrease in iron, actinolite becomes white and thereby grades into tremolite. When compact and tough, either mineral is called nephrite, which is one of the true jade minerals. Actinolite is found in prismatic, monoclinic crystals in Austria and Vermont.

Adamantine spar. An old name for a grayish-brown corundum from India which shows a bright luster on the parting surfaces.

Adelite group. A group of arsenate minerals that crystallize in the orthorhombic system and have the same atomic structure. It includes the following members, several of which form a partial series:

Adelite	$CaMg(AsO_4)(OH,F)$
Conichalcite	$CaCu(AsO_4)(OH)$
Austinite	$CaZn(AsO_4)(OH)$
Duftite	$PbCu(AsO_4)(OH)$

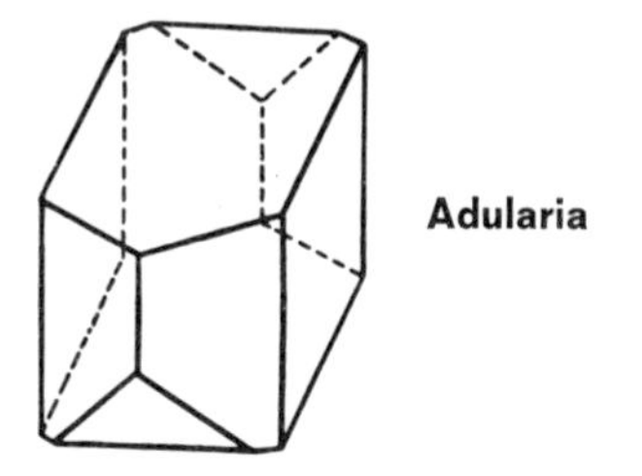
Adularia

Adularia. The low-temperature variety of orthoclase feldspar. It is colorless and more or less clear. Some of it makes up the gem moonstone.

Aegirite. A member of the spodumene series of minerals in the pyroxene family. It is a silicate of sodium and iron, $NaFe(Si_2O_6)$. Its luster is vitreous and the color is dark green or brown. Arkansas yields good crystals, but aegirite is known chiefly as a rock-forming mineral in certain rare, soda-rich rocks in Norway, Greenland, the Soviet Union, and Montana. Aegirine is the British spelling. Acmite is a special variety.

Aethiops mineral: An old name for metacinnabar. The origin of the word is obscure.

Agalmatolite. A compact variety of the mineral pyrophyllite or pinite, or some other silicate mineral or mixture or alteration product. It has been much used for Chinese carvings that resemble jade.

Agate. Chalcedony showing parallel bands of color. These are usually wavy or curved, although some are circular. According to the many variations in patterns, names such as *eye agate,* *fortification* (or *ruin*) *agate, banded agate,* and *iris agate* are used. The United States (especially the West), Uruguay, and Brazil are the main sources of agate. Some so-called agate, such as moss agate, is a general variety of chalcedony that has the colors distributed in irregular patches. Onyx and sardonyx are true agate.

Agatized wood. A petrified wood, filled or replaced by chalcedony that shows swirling patterns or patches of color.

Alabandite. An isometric mineral composed of manganese sulfide, MnS. It is black and nearly metallic looking. Arizona, Colorado, and Rumania are good localities.

Alabaster. The compact variety of gypsum, suitable for carving. Ancient alabaster was the mineral calcite of the variety called *onyx marble*. Gypsum alabaster is white or delicately tinted brown; it is mined in Italy and Colorado.

Alaska diamond. A wrong, commercial name for black hematite. This mineral is also sold as "Alaska black diamond," or under local names, for tourist purchase.

Albite. A member of the plagioclase feldspar series of minerals. It consists of 90 to 100 percent albite; the rest is anorthite. Outstanding crystals occur in Switzerland and Austria. A platy variety is called cleavelandite. Peristerite is albite, and so is some moonstone.

Alexandrite-sapphire. A gem variety of corundum having the color of alexandrite chrysoberyl.

Alkali-amphibole series. A series of monoclinic, silicate minerals in the amphibole family. It is also known as the glaucophane-riebeckite series and includes the following members:

Glaucophane	$Na_2Mg_3Al_2Si_8O_{22}(OH)_2$
Riebeckite	$Na_2Fe_3Fe_2Si_8O_{22}(OH)_2$

Allanite. A radioactive mineral found in granite and pegmatite. It is usually black and pitchy looking and occurs in irregular lumps, although its crystals are shaped like those of epidote. Allanite belongs to the epidote group and is a hydrous silicate of complex and variable composition; it includes enough thorium to cause weak radioactivity and a breakdown of the atomic structure. Its crystallization is monoclinic, and the formula is $(Ca,Ce,La,Na)_2(Al,Fe,Mn,Be,Mg)_3O(SiO_4)(Si_2O_7)(OH)$. The localities of interest include Sweden, Greenland, the Soviet Union, and Madagascar.

Allemontite. A hexagonal mineral composed of arsenic and antimony, AsSb. The metallic-gray masses occur in rounded shapes. The best locality is in France.

Aleppo stone. An old name for eye agate.

Allopalladium. A hexagonal mineral composed mostly of palladium, with other metals. It is metallic-white to gray and very heavy. It occurs in Germany and South Africa.

Allophane. An amorphous clay mineral.

Alluaudite series. A series of phosphate minerals which grade into each other and have similar physical properties, as well as a similar origin. It includes the following members:

Alluaudite	$(Na,Fe,Mn)PO_4$
Mangan-alluaudite	$(Na,Mn,Fe)PO_4$

Almandine-spinel. A gem variety of spinel having a violet color that resembles almandite garnet.

Almandite. A subspecies of the garnet group of minerals, being a silicate of iron and aluminum, $Fe_3Al_2(SiO_4)_3$. When deep red it is one of the so-called precious garnets; when reddish brown it is one of the so-called common garnets. Good crystals come from Alaska, and gem material occurs in India, Ceylon, and Brazil. Almandite is a typical metamorphic mineral and is the garnet usually used as an abrasive. New York is a leading source of such industrial material. As the variety rhodolite, almandite grades into pyrope.

Altaite. An isometric mineral composed of lead telluride, PbTe. It occurs in heavy, metallic-white masses in the United States, British Columbia (Canada), Chile, Western Australia, and Siberia.

Alum group. A group of sulfate minerals that crystallize in the isometric system and have the same atomic structure and a similar origin and occurrence. It includes the following members:

Potash alum	$KAl(SO_4)_2 \cdot 12H_2O$
Soda alum	$NaAl(SO_4)_2 \cdot 12H_2O$
Ammonium alum	$(NH_4)Al(SO_4) \cdot 12H_2O$

Alumogel. A name once given to certain constituents of the rocks bauxite and laterite, presumed to originate as gels.

Alumstone. A popular name for the mineral alunite, because of its chemical composition.

Alunite. A hexagonal, sulfate mineral, $KAl_3(OH)_6(SO_4)_2$, usually found as white, gray, or red masses. Well-known localities are in Italy, Hungary, Spain, New South Wales (Australia),

Colorado, Nevada, and Utah. Alunite is a source of alum and is sometimes an ore of potassium and aluminum.

Alunite group. A group of sulfate minerals that crystallize in the hexagonal system and have the same atomic structure. They grade rather freely into one another to make several natural series. The members are as follows:

Alunite series	
Jarosite	$KFe_3(SO_4)_2(OH)_6$
Ammoniojarosite	$(NH_4)Fe_3(SO_4)_2(OH)_6$
Natrojarosite	$NaFe_3(SO_4)_2(OH)_6$
Argentojarosite	$AgFe_3(SO_4)_2(OH)_6$
Carphosiderite	$[(H_2O)Fe_3(SO_4)_2(OH)_5 \cdot H_2O)]$
Beaverite	$Pb(Cu,Fe,Al)_3(SO_4)_2(OH)_6$
Plumbojarosite	$PbFe_6(SO_4)_4(OH)_{12}$

Alunite series. A series of hexagonal, sulfate minerals in the alunite group, which grade completely into each other and have similar physical properties. It includes the following members:

Alunite	$KAl_3(SO_4)_2(OH)_6$
Natroalunite	$NaAl_3(SO_4)_2(OH)_6$

Amalgam. A natural alloy of mercury and silver. This mineral is silver colored.

Amatrix. A popular name for a gemmy mixture of variscite and its matrix, which consists of quartz and chalcedony quartz. The name is derived from the words "American matrix."

Amazonite. A green variety of microcline feldspar. It is used as a gem and comes from Colorado, Virginia, the Ural Mountains, Norway, and Madagascar. It is also known as amazonstone.

Amazon jade. A wrong name for the green, amazonite variety of microcline feldspar.

Insect in amber

Amber. A fossil resin of ancient trees, used as a gem. It oozed from prehistoric pine trees, trapping insects and other organic objects in its sticky substance. The color varies, and several varieties have a somewhat different chemical composition. The Baltic coast is the principal locality; other amber comes from Sicily, Burma, and Rumania.

Amblygonite. A member of the amblygonite series of minerals. It occurs mainly as white masses of a vitreous or pearly luster. The occurrence is in pegmatite in South Dakota, California, and elsewhere. The crystallization is triclinic, and the formula is $(Li,Na)Al(PO_4)(F,OH)$.

Amblygonite series. A series of triclinic, phosphate minerals, which grade almost completely into one another and have similar physical properties, as well as a similar origin and occurrence. It includes the following members:

Amblygonite	$(Li,Na)Al(PO_4)(F,OH)$
Montebrasite	$(Li,Na)Al(PO_4)(OH,F)$
Natromontebrasite	$(Na,Li)Al(PO_4)(OH,F)$

Amethyst. Purple or violet quartz. This is the most valuable variety of quartz. The color, presumably due to iron, occurs in patches and layers. Uruguay, Brazil, Siberia, Ceylon, and Japan are among the leading sources.

Amosite. A long-fibered asbestos variety of anthophyllite.

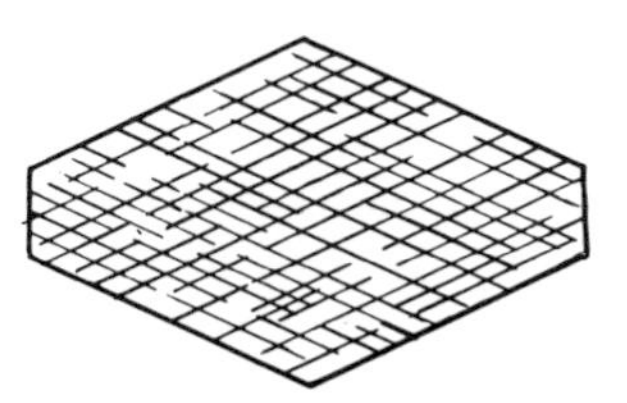

Amphibole cleavage

Amphibole family. A major family of rock-forming minerals. They are silicates of various metals, of which hornblende has the most complex combination and is the most common. The amphiboles may be placed in the anthophyllite, cummingtonite, tremolite-actinolite, hornblende, and alkali-amphibole series. There is a strong resemblance between many of the amphibole minerals and those in the related pyroxene family. Amphibole, however, has a large-angled cleavage, no parting, and a greater tendency toward elongated crystals. Amphibole forms at a lower temperature than pyroxene and is hydrous.

Analcime. The most distinctive member of the zeolite family of minerals. Its isometric crystals (trapezohedrons) are colorless or white, have a vitreous luster, and grow free in cavities, usually in lava. Other occurrences are known but this one is typical. The formula is $Na(AlSi_2O_6)H_2O$.

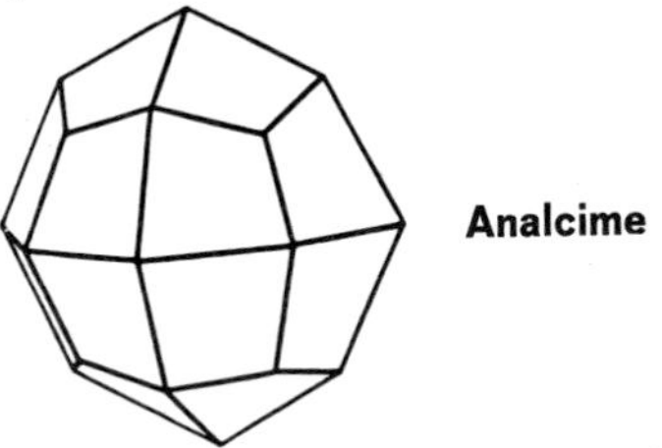

Analcime

Anatase. A tetragonal mineral composed of titanium oxide, TiO_2. Also known as octahedrite, it is a brown mineral with a bright, nonmetallic luster. Formed in veins and crevices, it is often found in sand. The Alps, Great Britain, the Soviet Union, and Brazil are important sources.

Andalusite. An industrial and gem mineral. It is an aluminum silicate, $AlAlO(SiO_4)$, occurring usually in coarse, nearly square, orthorhombic prisms of considerable hardness (7½). These vitreous crystals sometimes contain a dark cross of carbonaceous material that characterizes the variety called chiastolite. Andalusite, a metamorphic mineral, is found in Spain, Austria, and Brazil, and is mined as a refractory in California and the Transvaal (South Africa). Viridine is a green variety.

Andesine. A member of the plagioclase feldspar series of minerals. It consists of 50 to 70 percent albite and the rest anorthite and is common in the Andes Mountains.

Andorite group. A group of sulfosalt minerals that crystallize in the orthorhombic system and are probably similar in atomic structure and physical properties. It includes the following members:

Andorite	$PbAgSb_3S_6$
Lindstromite	$PbCuBi_3S_6$

Andradite. A subspecies of the garnet group of minerals. It is a silicate of calcium and iron, $Ca_3Fe_2(SiO_4)_3$. Besides the usual yellow and brown colors of garnet, andradite may also be green (called demantoid) or black (melanite). Andradite occurs in ore deposits in marble.

Anglesite. A heavy mineral of the barite group, $PbSO_4$. The luster ranges from adamantine to earthy, and the orthorhombic crystals are varied in shape. Anglesite is found in lead deposits, often with galena. Among the important sources are Sardinia, Wales, Scotland, South-West Africa, and New South Wales (Australia).

Anhydrite. An orthorhombic, sulfate mineral, $CaSO_4$. It is most typically found in white, blocky masses, although it may assume various tinges of color. It is often associated with gypsum and rock salt. Outstanding localities include Nova Scotia, New Mexico, Texas, Austria, Poland, and Germany.

Ankerite. A mineral of the dolomite group, $Ca(Fe,Mg,Mn)(CO_3)_2$. It occurs in central Europe and England.

Annabergite. A member of the erythrite-annabergite series of minerals in the vivianite group. It alters readily from niccolite and occurs as light-green crusts. Germany, Ontario, and Nevada are leading sources. The crystallization is monoclinic, and the formula is $(Ni,Co)_3(AsO_4)_2 \cdot 8H_2O$.

Anorthite. A member of the plagioclase feldspar series of minerals. It consists of 0 to 10 percent albite and the rest anorthite.

Anorthoclase. A variety of microcline feldspar, $(Na,K)AlSi_3O_8$, in which the sodium content exceeds the potassium.

Anthophyllite. A member of the anthophyllite series of minerals in the amphibole family. A hydrous silicate of magnesium and iron, $(Mg,Fe)_7(Si_8O_{22})(OH)_2$, it is vitreous and green or brown. A variety called amosite is a long-fibered asbestos. Greenland is an outstanding source of anthophyllite, which also occurs in metamorphic rocks in Norway, Pennsylvania, North Carolina, and Montana.

Anthophyllite series. A series of orthorhombic, silicate minerals in the amphibole family. It includes the following members:

Anthophyllite	$(Mg,Fe)_7Si_8O_{22}(OH)_2$
Gedrite	$(Mg,Fe,Al)_7(Si,Al)_8O_{22}(OH)_2$

Antigorite. A platy variety of serpentine.

Antimonial copper. An old name for chalcostibite, in reference to its chemical composition.

Antimonial lead ore. An old name for bournonite, referring to its chemical composition and use.

Antimonial red silver. An old name for pyrargyrite, because of its chemical composition and color.

Antimonial sulphuret of silver. An old name for freieslebenite, from its chemical composition.

Antimony blende. An old name for kermesite, from its chemical composition and bright but nonmetallic luster.

Antimony glance. An old name for stibnite, in reference to its chemical composition and bright, metallic luster.

Antimony ocher. An old name for cervantite, stibiconite, and perhaps other minerals containing antimony and having an earthy look.

Antlerite. A green mineral composed of hydrous copper sulfate, $Cu_3(OH_4)SO_4$. The slender crystals are vitreous, show striations, and occur in copper deposits in arid regions. Chile is by far the main source for this ore of copper.

Apache tears. A fanciful name given to rounded nodules of obsidian from the American West.

Apatite. An important phosphate mineral, $Ca_5(F,Cl,OH)(PO_4)_3$, of the apatite group. It occurs as *fluorapatite, chlorapatite, hydroxylapatite, carbonate-apatite,* and *collophane.* The hexagonal crystals are usually green or brown, although blue, violet, and colorless varieties are known. The luster is vitreous to resinous. Asparagus stone is a gem variety. The chief localities of apatite, except as phosphate rock, are in the Soviet Union, Norway, Sweden, Ontario, and Quebec.

Apatite group. A group of phosphate-arsenate-vanadate minerals and related minerals that crystallize in the hexagonal system and have the same atomic structure. It includes the following members:

Apatite series	
Pyromorphite series	
Svabite series	
Dehrnite	$(Ca,Na,K)_5(PO_4)_3(OH)$
Lewistonite	$(Ca,K,Na)_5(PO_4)_3(OH)$
Fermorite	$(Ca,Sr)_5(P,AsO_4)_3(F,OH)$
Wilkeite	$(Ca_5)(P,S,Si,CO_4)_3(OH)$
Ellestadite	$(Ca_5)(Si,S,P,CO_4)_3(Cl,F,OH)$

Apatite series. A series of phosphate minerals in the apatite group, which grade completely into one another and have similar physical properties. It includes the following members:

Fluorapatite	$Ca_5(PO_4)_3F$
Chlorapatite	$Ca_5(PO_4)_3Cl$
Hydroxylapatite	$Ca_5(PO_4)_3(OH)$
Carbonate-apatite	$Ca_{10}(PO_4)_6(CO_3)\cdot H_2O$

Apophyllite. A mineral found in cavities in lava rock. It is white, unless tinged by impurities, and it occurs in pearly and vitreous crystals of tetragonal form. The composition is a complex hydrous silicate, $KCa_4(Si_4O_{10})_2F \cdot 8H_2O$. Good localities are Germany, Czechoslovakia, Italy, India, Iceland, Greenland, Mexico, Nova Scotia, New Jersey, and Michigan.

Aquamarine. The blue, gem variety of the mineral beryl. It is vitreous, usually very clear, and may be found in large crystals. The color often tends toward green. Brazil, Siberia, Madagascar, and several parts of the United States yield good aquamarine.

Aragonite. A mineral of the aragonite group, $CaCO_3$. It occurs as vitreous crystals of several forms, and is usually white. Flos ferri is a distinctive variety; some onyx marble is also aragonite. Outstanding specimens come from Spain, Czechoslovakia, England, and the United States.

Aragonite group. A group of carbonate minerals that crystallize in the orthorhombic system and have the same atomic structure and similar physical properties. They grade somewhat into one another to make several natural series. The members are as follows:

Aragonite	$CaCO_3$
Witherite	$BaCO_3$
Strontianite	$SrCO_3$
Cerussite	$PbCO_3$

Arfvedsonite. A member of the amphibole family of minerals. It is often known as one of the soda amphiboles because of its sodium content. The formula is $Na_3Mg_4Al(Si_8O_{22})(OH,F)_2$.

Argentiferous sulphuret of copper. An old name for the mineral stromeyerite, in reference to its chemical composition.

Argentite. An isometric mineral composed of silver sulfide, Ag_2S. It is lead-gray (tarnishing black), heavy, and can be cut with a knife. Rich in silver, argentite is found in many silver mines, especially in the Western Hemsphere, but also in Norway and central Europe.

Argentite group. A group of sulfide-selenide-telluride minerals that crystallize in the isometric system and have similar atomic structures. It includes the following members:

Argentite	Ag_2S
Aguilarite	$Ag_2(Se,S)$
Naumannite	Ag_2Se
Digenite	$Cu_{2-x}S$
Berzelianite	Cu_2Se

Eucairite	CuAgSe
Hessite	Ag_2Te
Petzite	Ag_3AuTe_2

Argyrodite series. A series of sulfosalt minerals that crystallize in the isometric system and have the same atomic structure and similar physical properties. It includes the following members:

Argyrodite	Ag_8GeS_6
Canfieldite	Ag_8SnS_6

Arizona ruby. A wrong name for pyrope garnet, indicating one of its important sources and its color.

Arkansas diamond. A wrong name for rock-crystal quartz, referring to its important source and the bright luster.

Arrows of love. A popular name for rock-crystal quartz that encloses needlelike arrows of other minerals. *Sagenite* is the mineralogical name.

Arsenic group. A group of native-element minerals that crystallize in the hexagonal system and have the same atomic structure. It includes the following members:

Native arsenic	As
Allemontite	AsSb
Native antimony	Sb
Native bismuth	Bi

Arsenolite group. A group of oxide minerals that crystallize in the isometric system and have the same atomic structure but different physical properties. It includes the following members:

Arsenolite	As_2O_3
Senarmontite	Sb_2O_3

Arsenopyrite. A monoclinic mineral composed of iron sulfarsenide, FeAsS. It occurs in silver-colored crystals. It is widespread in South Dakota, Ontario, Bolivia, England, and Germany. It is a source of arsenic, which is recovered by smelting.

Arsenopyrite group. A group of sulfide minerals that crystallize in various systems and are related more in chemical composition than in atomic structure. The following members are included, though lautite is less certainly a member of this group:

Arsenopyrite	FeAsS
Glaucodot	(Co,Fe)AsS
Gudmanite	FeSbS
Lautite	CuAsS

Asbestos. Any of a number of fibrous, silicate minerals. The most important commercially is chrysotile serpentine; others include anthophyllite, crocidolite, and tremolite.

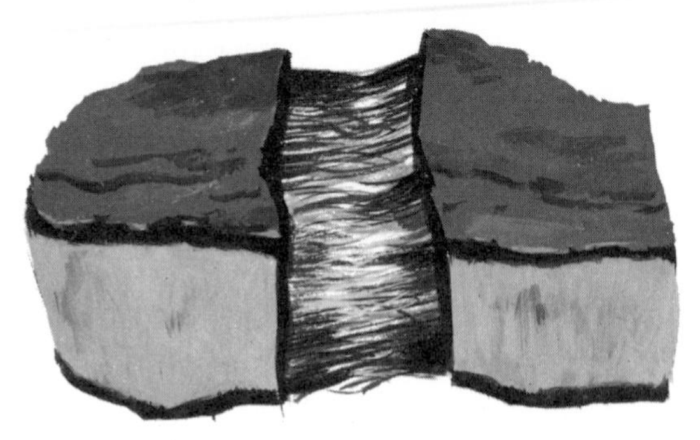
Asbestos

Asparagus stone. A transparent, green, gem variety of apatite. Mexico (Durango) is the chief source.

Atacamite. An orthorhombic mineral composed of copper chloride, $Cu_2(OH)_3Cl$. This green mineral is a minor ore of copper in Latin America (Chile, Bolivia, Mexico), the United States (Arizona, Utah), and South Australia.

Atacamite group. A group of halide minerals that crystallize in the orthorhombic system and have the same atomic structure. It includes the following members:

Atacamite	$Cu_2(OH)_3Cl$
Kempite	$Mn_2(OH)_3Cl$

Attapulgite group. A group of clay minerals having a rodlike shape. They are composed of hydrous silicates of aluminum and magnesium.

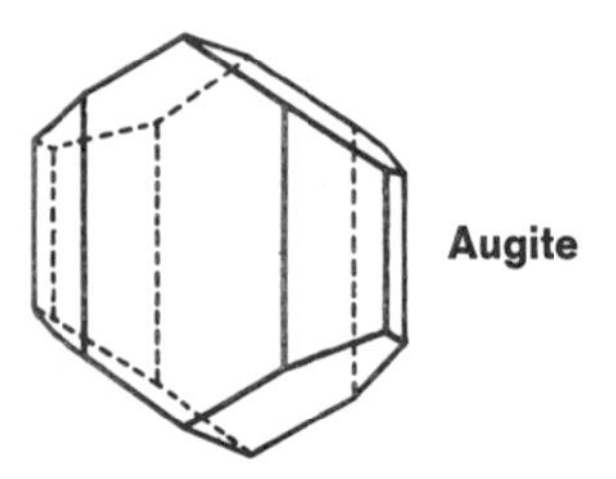
Augite

Augite. An important dark mineral in igneous rocks, and the most common member of the pyroxene family. It is a complex silicate and may be placed in the diopside-hedenbergite series.

Both the vitreous, monoclinic crystals and the masses of augite are dark green to black. Fine crystals are known from Italy, Czechoslovakia, and Mount Vesuvius.

Aurichalcite. An orthorhombic, carbonate mineral, $(Zn,Cu)_5(OH)_6(CO_3)_2$. It has a pale green to blue color and a silky to pearly luster. Arizona, Utah, and New Mexico yield fine specimens.

Aurosmiridium. An isometric mineral composed of iridium with dissolved osmium, ruthenium, and gold. It is hard, very heavy, and metallic white. It has come only from the Ural Mountains.

Autunite. A commercial, radioactive (uranium-bearing) mineral. It occurs in yellow to green crystalline plates and scales having a vitreous or pearly luster; under ultraviolet light it fluoresces yellowish green. The crystallization is tetragonal, and its formula is $Ca(UO_2)_2(PO_4)_2 \cdot 10\text{–}12H_2O$.

Aventurine. A spangled variety of quartz. The color is usually green, brown, or red, from the presence of flecks of mica and hematite. The Soviet Union and India are important sources. Aventurine feldspar is another name for sunstone.

Axstone. A popular name for nephrite jade, referring to the use of this mineral for axes and clubs by the Maoris in New Zealand.

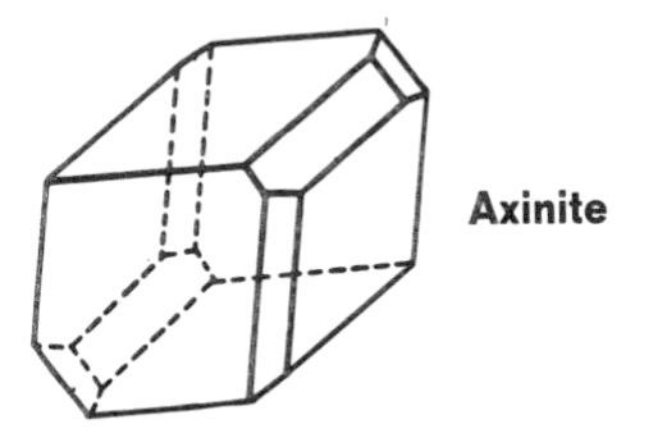

Axinite

Axinite. A triclinic mineral occurring in thin, sharp, wedge-shaped crystals—which are vitreous and mostly brown or violet. Good localities are in New Jersey, California, France, Switzerland, England, and Japan. The formula is $Ca_2(Fe,Mn)Al_2(BO_3)(Si_4O_{12})(OH)$.

Azurite. A copper mineral, $Cu_3(CO_3)_2(OH)_2$. It is blue and occurs in crystals and nodules. France, South-West Africa, New South Wales (Australia), and Arizona are the leading sources.

Azurmalachite. A gemmy mixture of azurite and malachite.

Baddeleyite. A monoclinic mineral composed of zirconium oxide, ZrO_2. It occurs in rounded crystals, often brown, in the gem gravels of Ceylon and the diamond sand of Brazil.

Balas ruby. A wrong name for pink and rose-colored varieties of the mineral spinel, from a source in Badakhshan, Afghanistan, and a resemblance to ruby.

Banded agate. Agate showing conspicuous bands of color. All true agate should have this pattern.

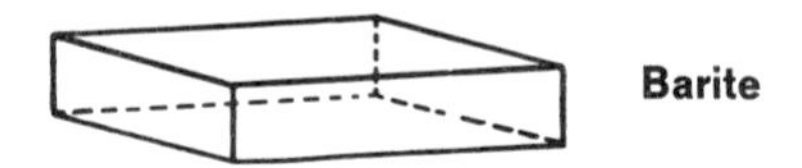

Barite

Barite. One of the most common minerals, a member of the barite group, $BaSO_4$. It occurs in good, often complex, orthorhombic crystals and tabular groups, which are vitreous and colorless, white, or blue. Its high density is noticeable. The tabular varieties are called *crested barite* and *barite roses.* Barite has numerous industrial uses; crushed barite is used as a tool-supporting sludge during deep well drilling. Important sources are in England, Rumania, Czechoslovakia, Germany, and Missouri.

Barite group. A group of sulfate minerals that crystallize in the orthorhombic system and have the same atomic structure and similar physical properties. It includes the following members, the first two of which form a probably complete series with each other:

Barite	$BaSO_4$
Celestite	$SrSO_4$
Anglesite	$PbSO_4$

Basanite. A fine-grained, black variety of jasper. It is used as a touchstone for testing precious metals according to their streak.

Bauxite. A field term for aluminum ores. The main minerals are gibbsite, boehmite, and diaspore, mixed together in various proportions in a rock that ranges in color from white to red. The knobby appearance of bauxite makes it one of the most distinctive of earth materials. It is mined in large quantities in the Guianas (Netherlands and British), Jamaica, Arkansas, France, Hungary, the Soviet Union, and Indonesia.

Bean ore. A popular name for a distinctive variety of rounded pieces of limonite (an iron ore).

Beegerite. A metallic-gray mineral composed of lead-bismuth sulfide, $Pb_6Bi_2S_9$. It is probably isometric; it has been found in Colorado and Siberia.

Bell metal ore. An old name for stannite, a minor ore of copper and tin—the main metals used to make bells.

Benitoite. A rare gem mineral of blue color, closely resembling sapphire. It is a silicate of barium and titanium, $BaTiSi_3O_9$. Associated with neptunite, it is found in San Benito County, California.

Berthierite. A metallic-gray mineral composed of iron-antimony sulfide, $FeSb_2S_4$. It is orthorhombic in crystallization. It occurs in central Europe, Latin America, New South Wales (Australia), and New Brunswick (Canada).

Beryl. A major gem mineral. Its varieties include such important gems as emerald and aquamarine, as well as heliodor, morganite, and golden beryl. The simple hexagonal crystals are often huge (up to 40 tons), and common beryl, the main source of the metal beryllium, has recently been discovered in commercial amounts in various small aggregates difficult to identify. The rough material typically has a bluish-green cast. Beryl is a silicate of beryllium and aluminum, $Be_3Al_2(Si_6O_{18})$. It is vitreous and hard (7½ to 8). Many localities are known, mostly in pegmatite, in which very large specimens have been found in Madagascar, Brazil, Maine, Colorado, South Dakota, and elsewhere.

Beryl group. A group of silicate minerals that have a similar atomic structure though different crystallization. It includes the following members:

Beryl	$Be_3Al_2(Si_6O_{18})$
Cordierite	$Mg_2Al_3(AlSi_5O_{18})$

Beryllonite. A rare, colorless, gem mineral found only in Maine. It has a bright luster. The formula is $NaBe(PO_4)$, and the crystallization is monoclinic.

Berzeliite series. A series of isometric, arsenate minerals which grade into each other and have related physical properties, as well as a similar origin and occurrence. It includes the following members:

Berzeliite	$(Mg,Mn)_2(Ca,Na)_3(AsO_4)_3$
Manganberzeliite	$(Mn,Mg)_2(Ca,Na)_3(AsO_4)_3$

Betafite series. A series of oxide minerals that crystallize in the isometric system and have the same atomic structure and similar physical properties, according to the chemical composition. It includes the following members:

Betafite	$(U,Ca)(Cb,Ta,Ti)_3O_6 \cdot nH_2O$
Djalmaite	$(U,Ca,Pb,Bi,Fe)(Ta,Cb,Ti,Zr)_3O_9 \cdot nH_2O$
Ampangabeite	$(Y,Er,U,Ca,Th)_2(Cb,Ta,Fe,Ti)_7O_{18} \cdot nH_2O$

Beudantite group. A group of sulfate minerals that grade partly into one another, crystallize in the hexagonal system, and perhaps have the same atomic structure. The following members are included:

Beudantite	$PbFe_3(AsO_4)(SO_4)(OH)_6$
Corkite	$PbFe_3(PO_4)(SO_4)(OH)_6$
Hinsdalite	$(Pb,Sr)Al_3(PO_4)(SO_4)(OH)_6$
Svanbergite	$SrAl_3(PO_4)(SO_4)(OH)_6$
Woodhouseite	$CaAl_3(PO_4)(SO_4)(OH)_6$

Bindheimite group. A group of antimonate minerals that crystallize in the isometric system and have the same atomic structure. It includes the following members, as well as monimolite, which may be intermediate in a partial series.

Bindheimite	$Pb_2Sb_2O_6(O,OH)$
Romeite	$(Ca,Fe,Mn,Na)_2(Sb,Ti)_2O_6(O,OH,F)$

Biotite. The black variety of mica, although the color varies somewhat. Iron and magnesium are present in addition to the fundamental mica formula, thus becoming $K(Mg,Fe)_3(AlSi_3O_{10})(OH)_2$. Mount Vesuvius is a good locality, but biotite is worldwide in distribution.

Bismuth glance. An old name for bismuthinite, from its chemical composition and bright, metallic luster.

Bismuthic gold. An old name for maldonite, alluding to its chemical composition. *Bismuth-gold* is another name.

Bismuthinite. An orthorhombic mineral composed of bismuth sulfide, Bi_2S_3. It is metallic-gray and heavy. Bolivia has important deposits of this minor ore of bismuth.

Bismuth ocher. An old name for bismite and other minerals containing bismuth and having an earthy look.

Bismuthotellurites. An old name for tetradymite, indicating its chemical composition.

Bisulphuret of iron. An old name for pyrite, referring to its chemical composition.

Bitter salts. A popular name for epsomite, because of its taste.

Bixbyite. A black, metallic mineral composed of iron-manganese oxide, $(Mn,Fe)_2O_3$. It occurs in Utah, Argentina, Spain, India, Sweden, and South Africa.

Black-band ore. A miners' name for siderite (an iron ore) containing bituminous matter arranged in streaks.

Black cobalt ocher. An old name for wad containing cobalt, because of its black, earthy appearance.

Black copper. An old name for tenorite, from its color and metal content.

Black gold. An old name for the mineral maldonite, referring to its color (when tarnished) and metal content.

Black hematite. An old name for psilomelane, in reference to its resemblance to hematite.

Black iron ore. Another old name for psilomelane, indicating its resemblance to hematite.

Black jack. An English and American miners' name for dark varieties of the mineral sphalerite containing much iron.

Black lead. An old name for graphite, because of its appearance. Graphite is still called "lead" when used in pencils.

Black manganese. An old name for hausmannite, because of its color and chemical composition.

Black opal. Precious opal having a dark background; this is usually dark blue or gray, against which the play of colors flashes with spectacular effect. Some of the material, known as opal-matrix, must be cut with its native rock included. New South Wales and Queensland (Australia) and Nevada are the main producers.

Black oxide of copper. An old name for tenorite, because of its color and chemical composition.

Black sand. Heavy sand containing an appreciable amount of opaque, heavy minerals. These include hematite, ilmenite, magnetite, and chromite. They occur with the usual constituents of sand (particularly quartz), as well as with transparent, heavy minerals, such as rutile, monazite, garnet, zircon, staurolite.

Black tellurium. An old name for nagyagite, in reference to its color and chemical composition.

Blende. A miners' name for sphalerite, because of its "blind" (misleading and nonproductive) resemblance to galena. The word was also used in combination to describe certain minerals having a bright but nonmetallic luster; examples include *antimony blende* (kermesite), *cadmium blende* (greenockite), and *fireblende* (pyrostilpnite).

Bloedite group. A group of sulfate minerals that crystallize in the monoclinic system and have the same atomic structure and a similar origin and occurrence. It includes the following members:

Bloedite	$Na_2Mg(SO_4)_2 \cdot 4H_2O$
Leonite	$K_2Mg(SO_4)_2 \cdot 4H_2O$

Bloodstone. Green chalcedony quartz with red spots. In Great Britain, it is called heliotrope. India and the Ural Mountains are the chief sources.

Blue asbestos. The blue, fibrous crocidolite.

Blue copper. An old name for covellite, referring to its color and metal content.

Blue copper carbonate. The mineral azurite.

Blue iron earth. An old name for vivianite, referring to its appearance and chemical composition.

Blue john. A popular name for the massive, blue, fibrous or finely granular varieties of fluorite from Derbyshire, England.

Blue stone. An old name for chalcanthite, referring to its color.

Blue vitriol. An old name for chalcanthite, alluding to its color and sulfate composition.

Blushing copper. A popular name for bornite, referring to its easy tarnish.

Boehmite. An orthorhombic mineral composed of aluminum hydroxide, AlO(OH). It is an important constituent of bauxite, the ore of aluminum, but is too fine grained to be identified at sight.

Bog iron ore. A miners' name for goethite and limonite, because of their common occurrence together in bogs and swamps. They cannot be separated on the basis of locality.

Bog manganese. Another name for wad, from its occurrence in swamps.

Bog ore. Another name for limonite and goethite, indicating a common occurrence for these iron ores.

Bolognian spar. An old name for the barite found near Bologna, Italy, referring to its bright luster on the cleavage surface.

Bone-phosphate. An old name for apatite, referring to its being a chief constituent of bones.

Bone-turquoise. Another name for odontolite, because of its occurrence as fossil bones and teeth.

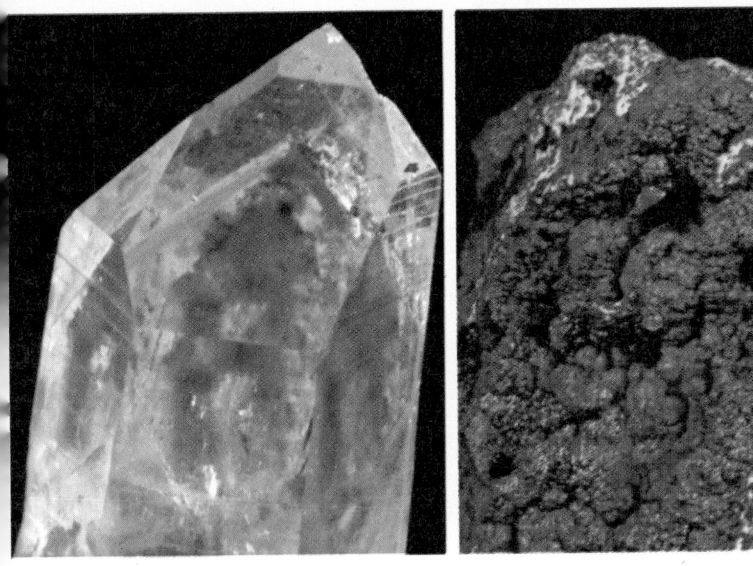

Quartz. Besides the mossy inclusions, a *phantom* quartz crystal can be seen within. Note also the sharp termination and the striations. From Hot Springs, Arkansas.

Chalcophanite. This is the usual form of this uncommon mineral from Mexico.

Crazy-lace agate is a very popular ornamental stone. From Mexico.

Agate. Showing perfect, parallel banding, this specimen is typical of the fine agates from Winona, Minnesota.

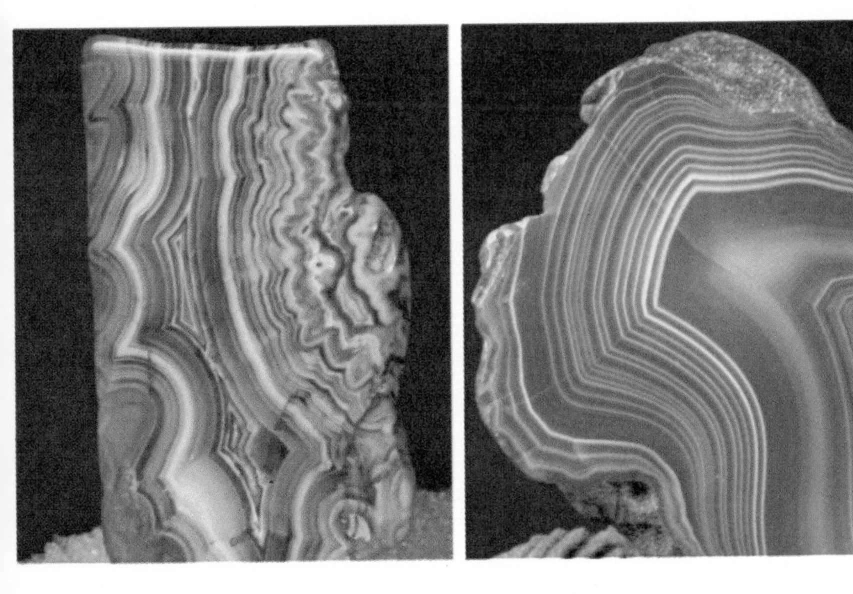

Amethyst (lavender quartz), citrine (yellow quartz), and rock crystal (clear quartz) with calcite (frosty spikes). A close-up of the intimate realms of the crystal world. The quartzes, commonest of minerals, include more gem varieties than any other family. This cluster is from Mexico.

Magnetite. The dodecahedrons of this magnetic iron ore are visible here. The yellow coating is limonite, an alteration product of the magnetite. From Arizona.

Stibiconite from Mexico. The crystal shafts suggest the form of stibnite, from which this oxide of antimony has altered. This mineral is rarely found in crystals.

Quartz geode. This specimen, from Hamilton, Illinois, has a curious inner growth of quartz crystals.

Hematite needle ore. The lustrous appearance is typical. The locality is Ishpeming, Michigan, in the iron ranges of the Lake Superior region.

Goethite. This fibrous variety is from Ishpeming, Michigan. The satiny appearance of this mineral, named after the German poet, is due to its fibrous structure.

Fergusonite. This odd-looking pair has a bright luster where fractured. This specimen came from near Ambatofotsikely, Madagascar.

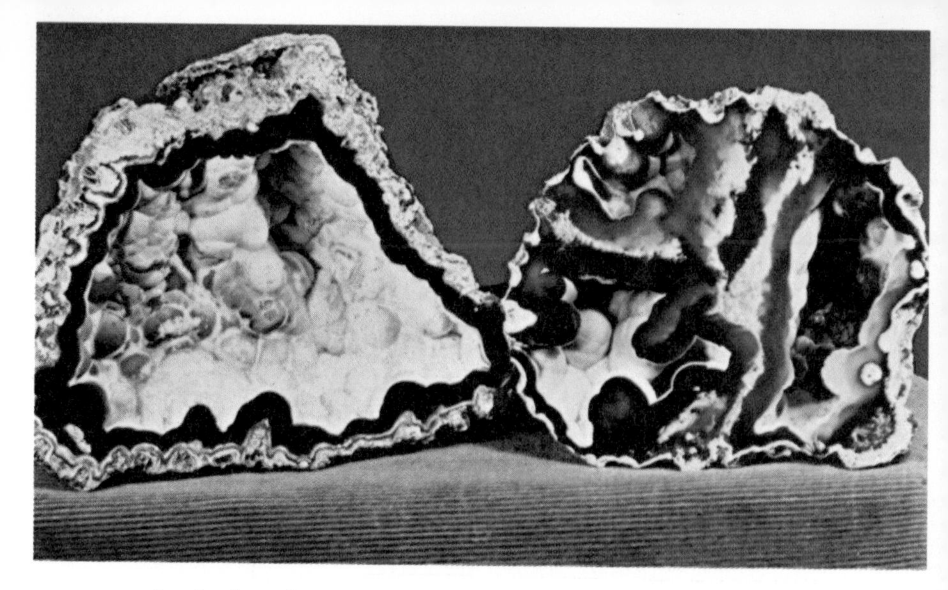

Agatized coral. The rough and barnaclelike crust of these nodules, a pseudomorph of chalcedony after coral, belies their attractive interior. From Tampa, Florida.

Quartz pseudomorphs after aragonite. The long, sugary-looking crystals were once aragonite, but quartz has replaced it. From Austria.

Cuprite, chrysocolla, and minium. This parti-colored rock from Arizona has a center of cuprite, a rim of chrysocolla, and ribbons of bright minium.

Fluorite with quartz. A contrast in color and form is this specimen from Rosiclare, Illinois. The large fluorite cube combined with the sparkling quartz clusters make this a collectors' delight.

Paralaurianite and cerussite. These are intimately associated in this frosty flower from Tiger, Arizona.

CARBONATES

Rhodocrosite crystals with goethite. In this close-up, the crystals of golden rhodocrosite, which look like barley-sugar candy, contrast with the shiny, black bubbles of goethite. This fine grouping is from Germany. Carbonates include many of the most decorative minerals —note especially the calcite, malachite and azurite on the following pages.

Calcite twin.

This perfect *contact twin* from Baxter Springs, Kansas, shows a twinned form common in calcite—crystals formed side by side like Siamese twins.

Aragonite.

Flos ferri is an older name given to this exotic, coral-like form of aragonite. The brown stains derive from its occurrence with iron-ore deposits. From Bisbee, Arizona.

Cerussite, wulfenite, dioptase.

A marvelously clean specimen, showing distinct crystals of white cerussite complemented by the superb, green dioptase and yellow wulfenite. This handsome group is from Tiger, Arizona.

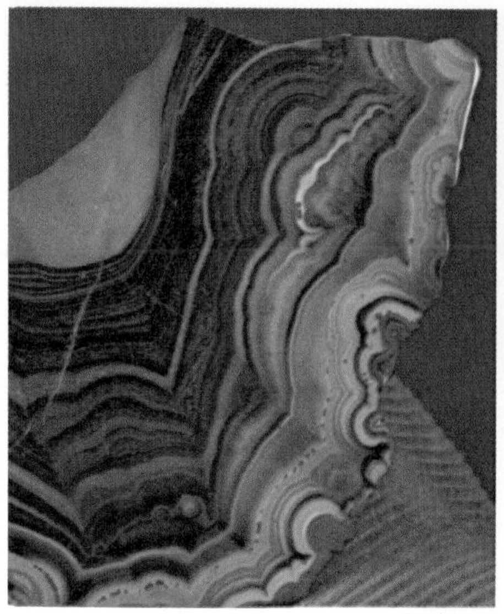

Banded malachite. This is the classic, and now seldom seen, ornamental variety from the Ural Mountains in the Soviet Union. The agatelike banding makes it almost a gem.

Icicle calcite. An ice-clear variety of this familiar mineral. This stairstep crystal is from Mexico.

Aragonite in one of its most pleasing aspects. This frosty-looking specimen is from Bisbee, Arizona.

Calcite. Like rockets pointed outward toward space are these well-terminated shafts of calcite from Cumberland, England.

Nicholsonite. A zinc-bearing variety of aragonite in tapered crystals with domed terminations; from Tsumeb, South-West Africa.

Dolomite crystals in geode. The lining is crystallized quartz. From Hamilton, Illinois.

Witherite. A pinnacle of uniformly sized crystals from England, all of them being *repeated twins,* the customary habit of witherite.

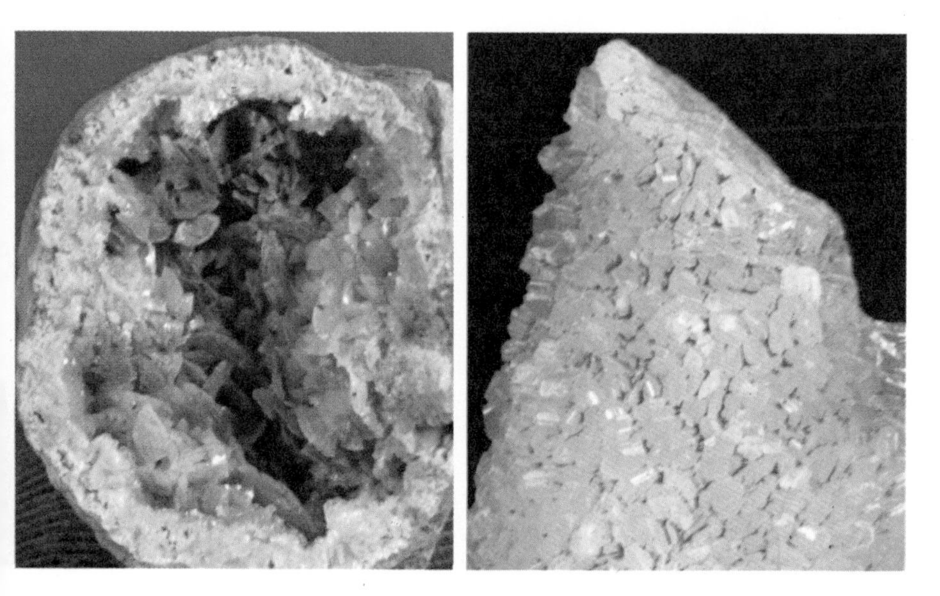

Calcite on native copper.

From Keweenaw County, Michigan, comes this jewel in a rough setting. The bright, golden flecks glinting from the interior of the calcite are inclusions of copper, untouched by the atmosphere that has tarnished the exposed copper.

Ankerite.

The creamy-white, sharp rhombs of this mineral show vividly. From Gilman, Colorado.

Malachite.

Masses of rich-green, needlelike crystals depict this popular copper carbonate mineral at its finest. Note the velvety tufts of minute crystals. From Mexico.

Cerussite and malachite. This mineral association of two carbonate minerals is from Tiger, Arizona. Cerussite is the dominant mineral.

Ankerite and sphalerite. Ankerite and black sphalerite in striking contrast. This specimen came from Gilman, Colorado, where the sphalerite is mined for zinc.

Azurite and malachite. A pair of closely related copper minerals. The azurite is the less common of the two carbonates. From the copper deposits of Bisbee, Arizona, comes this exquisite specimen.

Sand crystal.

Resembling a monolithic sculpture or an ancient artifact, this interesting crystal is quartz sand intermixed with calcite but retaining the general shape of a calcite crystal. This mineral oddity is from Rattlesnake Butte, South Dakota

Smithsonite.

A sparkling, emerald green makes this smithsonite specimen from Kelly, New Mexico, a very choice one.

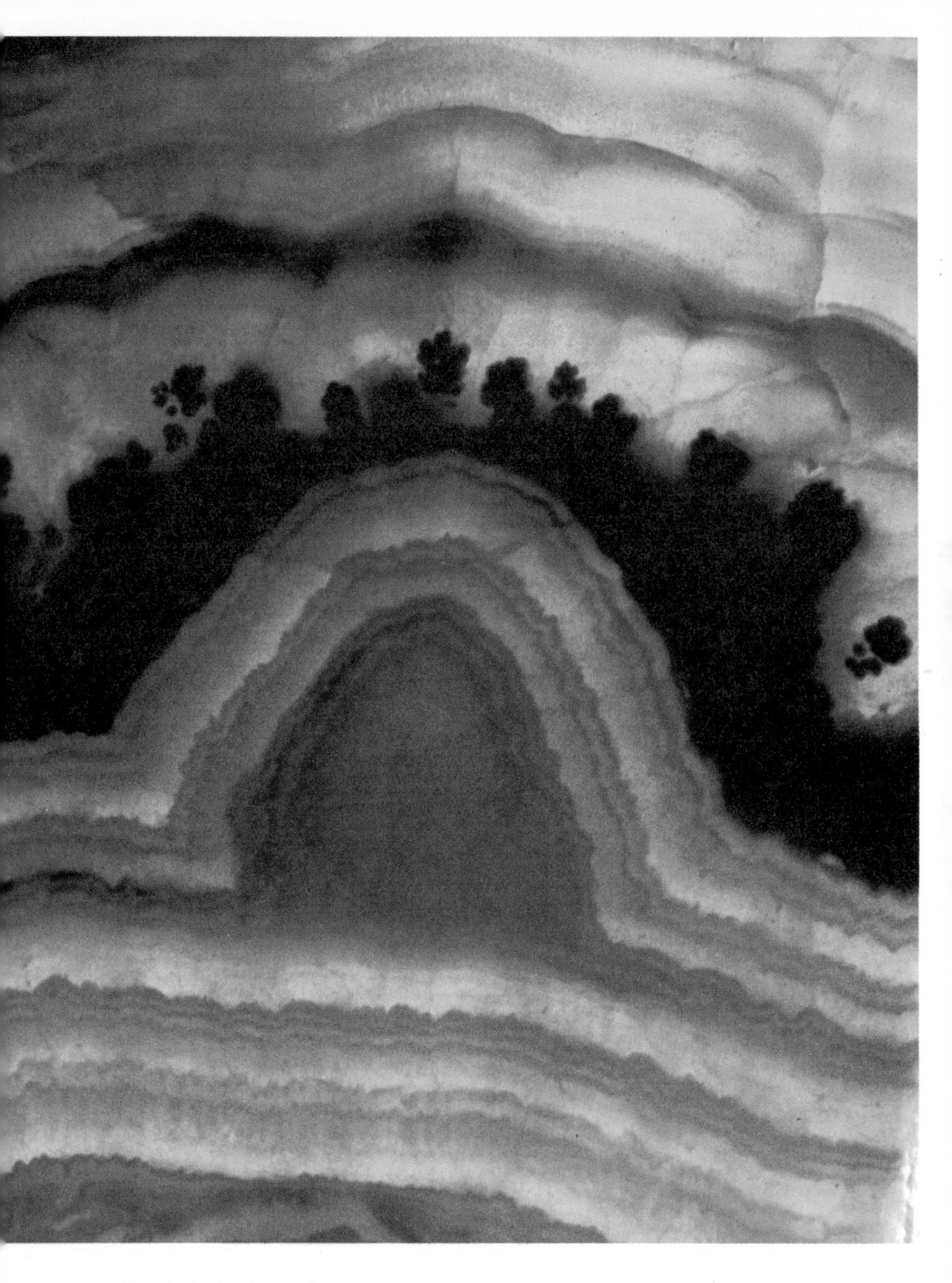

Banded rhodocrosite. This translucent slice reveals dendritic advances of manganese oxide rimming the flaming lower part. Much can be learned about crystallization from this specimen from Patagonia.

Azurite twin with malachite from Tsumeb, South-West Africa. The solid structure of azurite and the fibrous nature of malachite are clearly shown here. So mirrorlike is the surface of the azurite that it reflects the matrix material.

Aurichalcite with plattnerite. From Mexico comes this array of color, form, and fragility. Many of the blue aurichalcite needles are tipped with the iron-black plattnerite.

Boracite. An orthorhombic, borate mineral, $Mg_3B_7O_{13}Cl$. The crystals are hard, vitreous, and usually colorless. A product of evaporation, boracite is found in Germany and Louisiana.

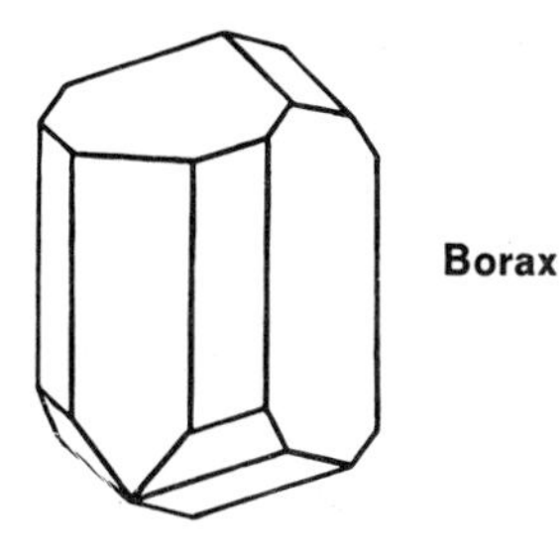

Borax

Borax. Natural borax in mineral form. The monoclinic crystals are vitreous and colorless; white masses are also found. Important sources include Tibet, Italy, California, and Nevada. The formula is $Na_2B_4O_7 \cdot 10H_2O$.

Bornite. An isometric mineral composed of copper-iron sulfide, Cu_5FeS_4. It has a brownish-bronze color when fresh but tarnishes so readily that the purple-blue-black, iridescent surface is usually the only one seen. Bornite is mined widely around the world for its copper content. It occurs in irregular masses with other copper minerals.

Borspar. A coined name for colemanite, from its boron content and bright luster on the cleavage surface.

Bort. A variety of industrial diamond. It is an aggregate of crystalline growths, which give it the needed toughness. The Congo is the chief source.

Boulangerite. A monoclinic mineral composed of lead-antimony sulfide, $Pb_5Sb_4S_{11}$. Having a fibrous look and a metallic-gray color, it is one of the feather ores. It is found in veins of lead and zinc minerals. France and Peru have yielded substantial amounts.

Bournonite. An orthorhombic mineral composed of lead-copper-antimony sulfide, $PbCuSbS_3$. Its metallic-gray crystals are often twinned in a distinctive cogwheel shape. All three metals are obtained from it commercially. Outstanding localities for specimens include Germany, Rumania, and England.

Bournonite group. A group of sulfosalt minerals that crystallize in the orthorhombic system and have the same or a related crystal structure. It includes the following members:

Bournonite	$PbCuSbS_3$
Seligmannite	$PbCuAsS_3$
Aikinite	$PbCuBiS_3$

Bowenite. A variety of serpentine used as a substitute for jade. It is light yellowish green and comes from India, China, and New Zealand.

Braunite. A tetragonal mineral composed of manganese-silicon oxide, $(Mn,Si)_2O_3$. Its usual occurrence is in brownish-black masses, which are often used as an ore of manganese. Germany has several important localities for braunite.

Brazilian. A name added to several gems to suggest a more valuable gem. Examples include *Brazilian emerald* for green tourmaline, *Brazilian peridot* for yellowish-green tourmaline, and *Brazilian sapphire* for blue tourmaline.

Brazilianite. A gem mineral of rather recent discovery. It is a monoclinic phosphate, $NaAl_3(PO_4)_2(OH)_4$. Brazil and New Hampshire are the localities.

Breithauptite. A hexagonal mineral composed of nickel antimonide, NiSb. Violet-red specimens of this heavy mineral occur in Ontario (Cobalt) and in Germany.

Bright white cobalt. An old name for cobaltite, indicating its appearance and metal content.

Brittle feather ore. A miners' name for jamesonite and capillary stibnite, from their brittleness (due to cross cleavage) and feathery appearance. The rest of the feather ores are known as *flexible feather ores.*

Brittle mica group. A group of micalike minerals occurring in brittle flakes. It includes margarite, xanthophyllite, ottrelite, and chloritoid.

Brittle silver glance, Brittle silver ore, Brittle sulphuret of silver. Old names for stephanite, indicating its brittleness, chemical composition, and bright luster.

Brochantite. A green mineral composed of hydrous copper sulfate, $Cu_3(OH_4)SO_4$. It occurs in copper deposits as slender, vitreous crystals showing striations.

Bromyrite. An isometric mineral composed of silver bromide, AgBr. This hornlike, brown mineral can be cut like plastic. It forms a complete series with cerargyrite. Bromyrite is especially abundant in Chile.

Bronzite. An intermediate member of the enstatite-hypersthene series of minerals in the pyroxene family. Unlike the others, it has a distinctive bronzy luster.

Brookite. An orthorhombic mineral composed of titanium oxide, TiO_2. It is brown and has a bright luster. It is formed in veins and often erodes to become sand. Noted localities are in the Alps, Great Britain, the Soviet Union, and Brazil.

Brown clay ironstone. A familiar name for goethite, limonite, and siderite, referring to their color and composition.

Brown hematite. An old name for limonite and goethite, because of their color and iron content.

Brown iron-ore. A popular name for goethite and limonite, indicating their color and use.

Brown iron-stone. An old name for limonite and goethite, because of their color and chemical composition.

Brown ocher. An old name for limonite, indicating its color and earthy look.

Brown spar. A miners' name for brown magnesite, siderite, and brown dolomite, because of their color and bright luster on the cleavage surface.

Brucite. A hexagonal mineral composed of magnesium hydroxide, $Mg(OH)_2$. It occurs in light-colored, flexible leaves. Brucite is used for metallurgical purposes. Nevada is a leading source.

Brucite group. A group of hydroxide minerals that crystallize in the hexagonal system and have the same atomic structure and similar physical properties. It includes the following members:

Brucite	$Mg(OH)_2$
Pyrochroite	$Mn(OH)_2$
Portlandite	$Ca(OH)_2$

Brushite group. A group of phosphate-arsenate minerals that crystallize in the monoclinic system and have the same atomic structure. It includes the following members:

Brushite	$CaH(PO_4)\cdot 2H_2O$
Pharmacolite	$CaH(AsO_4)\cdot 2H_2O$

Bustamite. A member of the pyroxenoid group of minerals. It is a silicate of calcium and manganese, $CaMnSi_2O_6$.

Bytownite. A member of the plagioclase feldspar series of minerals. It consists of 10 to 30 percent albite and the rest anorthite.

Cacholong. A porous variety of common opal. It will adhere to the tongue.

Cadmium blende. An old name for greenockite, indicating its chemical composition and bright but nonmetallic luster.

Cairngorm. Yellow quartz, often smoky, grading into smoky quartz. Scotland was formerly the classic locality.

Calamine. An older and less acceptable name for the mineral hemimorphite. In Great Britain, calamine refers to the mineral called *smithsonite* in the United States.

Calaverite. A monoclinic mineral composed of gold telluride, $AuTe_2$. It is heavy, otherwise resembling silver more than gold, but it yields a button of gold when the tellurium is driven off by heat. Colorado, California, Ontario, Western Australia, and the Philippines are the chief sources of this ore of gold.

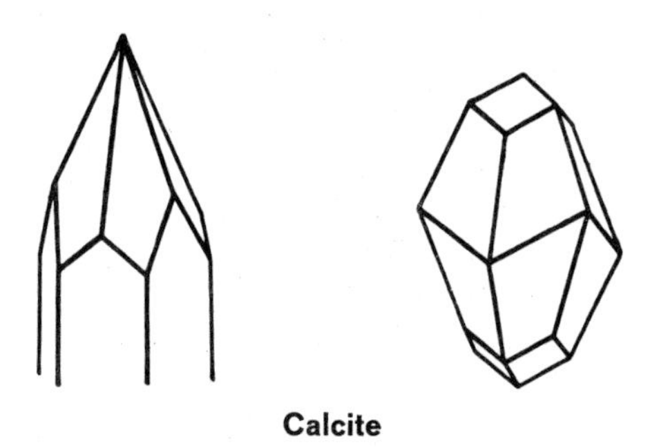

Calcite

Calcite. The most diversified of all known minerals, having been described in more than 300 different forms. The vitreous, light-colored hexagonal crystals are often very complex. Calcite also occurs as stalactites and compact and earthy masses. Among the familiar varieties of calcite are Iceland spar, sand crystal, travertine, and onyx marble. Calcite is of worldwide extent; the leading sources include localities in Germany, England, Iceland, Mexico, Missouri, and Michigan. It is a member of the calcite group; the formula is $CaCO_3$.

Calcite group. A group of carbonate minerals that crystallize in the hexagonal system and have the same atomic structure and similar physical properties. They grade into one another more or less readily. The members are as follows:

Calcite	$CaCO_3$
Magnesite	$MgCO_3$
Siderite	$FeCO_3$

Rhodochrosite	$MnCO_3$
Cobaltocalcite	$CoCO_3$
Smithsonite	$ZnCO_3$
Otavite	$CdCO_3$

Calc-sinter. An old name for travertine, indicating its calcium content and evaporite origin.

Calc spar. An old name for calcite, referring to its calcium content and bright luster on the cleavage surface.

California. A name added to several gems to suggest a more valuable gem. Examples include *California jade* for green idocrase and *California ruby* for red garnet.

Californite. A green, jadelike variety of idocrase. It comes from California and is used as a gem.

Calk, Cauk, Cawk. Derbyshire miners' name for barite.

Cancrinite. A hexagonal member of the feldspathoid group of minerals. It usually occurs as a yellow coating on nepheline, from which it has altered. The formula is $(Na_2Ca)_4(AlSiO_4)_6(CO_3,SO_4)\cdot 1\text{–}3H_2O$.

Cape ruby. A wrong name for pyrope garnet, indicating an important source (in South Africa) and its color.

Capillary pyrites. An old name for millerite, because it occurs in fibrous crystals having the color of pyrite.

Capillary red oxide of copper. An old name for the chalcotrichite variety of cuprite, because of its hairlike, red crystals and its chemical composition.

Carbonado. A variety of industrial diamond. It is opaque and dark, and tough enough for the most strenuous duties as a cutting and abrasive agent. It is found only in Brazil.

Carbonate-apatite. A variety of the mineral apatite, $Ca_{10}(PO_4)_6(CO_3)\cdot H_2O$.

Carbon group. A group of native-element minerals that have the same chemical composition but widely contrasting atomic structures and physical properties. It includes the following members:

Diamond C; Graphite C.

Carnallite An orthorhombic mineral composed of hydrous potassium-magnesium chloride, $KMgCl_3\cdot 6H_2O$. It is a white mineral with a bitter taste and occurs in granular masses. Carnallite is a source of potash and magnesium, and comes from Germany (Stassfurt) and the United States (New Mexico, Texas).

Carnelian. Red chalcedony quartz. It has long been a favorite gem for beads and small carvings. As it becomes brownish, it grades into sard.

Carnotite. An important radioactive mineral. It occurs as a yellow, earthy powder, often with petrified wood, especially in the western part of the United States. The formula is $K_2(UO_2)_2(VO_4)_2 \cdot 3H_2O$.

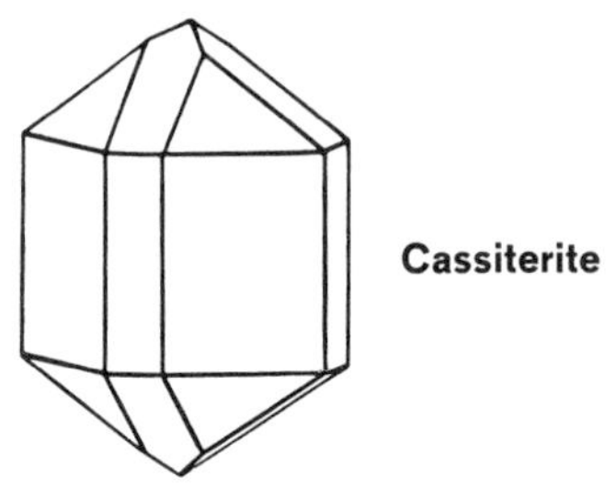

Cassiterite

Cassiterite. A commercially important mineral composed of tin oxide, SnO_2. It is usually massive and granular, often resembling wood or rocky pebbles. It is heavy and hard, and the color is usually brown or black; the crystals are tetragonal. As the chief ore of tin, cassiterite is recovered almost entirely from placer deposits in Malaya, Indonesia, the Congo, Thailand, and Nigeria. Some veins of it are mined in Bolivia.

Catlinite. A red, hardened clay. It is also known as *pipestone* and was used by the Indians for pipes. Minnesota and Wisconsin are the principal localities.

Cat's-eye quartz. A gem variety of quartz. It is less valuable than the true cat's-eye *(chrysoberyl).* Ceylon, India, and Germany are its sources.

Celadonite. A green mineral found in cavities in basalt. It is a complex hydrous silicate related to glauconite, perhaps $KMg_3Fe_3Si_9O_{25}(OH)_2 \cdot 9H_2O$.

Celestite. A mineral of the barite group, $SrSO_4$. The orthorhombic crystals are vitreous to pearly, and colorless, white, blue, or red. Sicily, England, Switzerland, and Ohio are major sources. It has several industrial uses, including beet-sugar refining and fireworks manufacturing

Cellular pyrites. An old name for marcasite, referring to its habit and resemblance to pyrite.

Celsian. A barium feldspar, $Ba(Al_2Si_2O_8)$.

Cerargyrite. An isometric mineral composed of silver chloride, $AgCl$. This unusual mineral typically has the appearance of gray horn or wax, and it can also be cut with a knife almost like plastic. It is an ore of silver and has come from some of the formerly great mines of Germany, Colorado, Nevada, and Idaho.

Cerargyrite series. A series of isometric, halide minerals, which grade completely into each other and have a similar origin and occurrence. It includes the following members:

Cerargyrite	$AgCl$
Bromyrite	$AgBr$

Cerussite. A mineral of the aragonite group, $PbCO_3$. It has a bright, nonmetallic luster and is usually white. Colorado (Leadville), Arizona, Idaho, New South Wales (Australia), Germany, and South-West Africa are among the leading localities. It is an ore of lead in many places.

Ceylon. A name added to several gems to suggest a more valuable gem. Examples include *Ceylon chrysolite* and *Ceylon peridot,* both applied to yellow-green tourmaline.

Ceylonite. A nearly black variety of spinel. It is colored by iron.

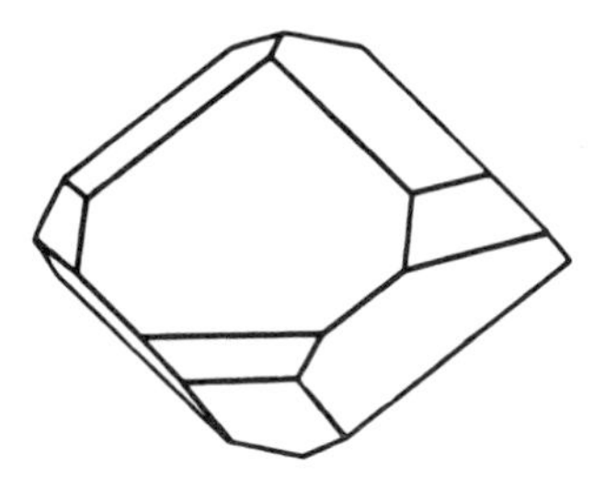

Chabazite

Chabazite. A member of the zeolite family of minerals. It forms vitreous, rhombohedral crystals which are either white, yellow, pink, or red. Its occurrence in certain volcanic-ash deposits may be rather common but it is not readily recognized. The formula is $(Ca,Na)_2(Al_2Si_4O_{12}) \cdot 6H_2O$.

Chalcanthite. A deep-blue mineral found in triclinic crystals and in masses, which are often fibrous. It has a vitreous luster and metallic taste. The occurrences are mostly in Chile. The formula is $CuSO_4 \cdot 5H_2O$.

Chalcanthite group. A group of sulfate minerals that probably have the same triclinic atomic structure and similar physical properties. It includes the following members:

Chalcanthite	$CuSO_4 \cdot 5H_2O$
Siderotil	$FeSO_4 \cdot 5H_2O$
Pentahydrite	$MgSO_4 \cdot 5H_2O$

Chalcedony. Cryptocrystalline quartz. It is flintlike in texture, which makes it especially suitable for cutting-tools and weapons. It comes in many colors and patterns, which give individual names to the varieties. These include carnelian, sard, prase, chrysoprase, plasma, agate, bloodstone, onyx, flint, chert, jasper.

Chalcocite. An orthorhombic mineral composed of copper sulfide, Cu_2S. It can be cut somewhat with a knife. It usually occurs in lead-gray masses. The western United States has many deposits of copper minerals with which chalcocite is found.

Chalcocite group. A group of sulfide minerals that crystallize in the orthorhombic system and have somewhat similar physical properties, although acanthite is still uncertain. The following members are included.

Chalcocite	Cu_2S
Stromeyerite	AgCuS
Acanthite	Ag_2S

Chalcopyrite. An important, isometric mineral composed of copper-iron sulfide, $CuFeS_2$. It is one of the fool's gold minerals, having a brassy color, which may tarnish. Its brittleness and intermediate hardness will distinguish it from the other fool's gold and from true gold. Chalcopyrite is the most important ore of copper and is also an important ore of gold. It is widely distributed throughout the world. Handsome masses come from Quebec (Rouyn).

Chalcopyrite group. A group of sulfide minerals that crystallize in the tetragonal system and have a similar atomic structure. It includes the following members:

Chalcopyrite	$CuFeS_2$
Stannite	Cu_2FeSnS_4

Chalcosiderite. A mineral of the turquoise group. It occurs as crusts and groups of vitreous, green crystals. The crystallization is triclinic, and the formula is $CuFe_6(PO_4)_4(OH)_8 \cdot 4H_2O$.

Chalcostibite group. A group of sulfosalt minerals that crystallize in the orthorhombic system and have similar atomic structures and physical properties. It includes the following members:

Chalopyrite	$CuSbS_2$
Emplectite	$CuBiS_2$

Chalcotrichite. The brilliant, red, fibrous variety of cuprite.

Chalybite. The British name for siderite.

Chamosite. An iron-bearing member of the chlorite group of minerals. It occurs with some sedimentary iron deposits.

Chert. A cryptocrystalline variety of quartz, occurring as a compact, massive rock similar to flint but lighter in color.

Chessylite. An old name for azurite, from its important occurrence at Chessy, France.

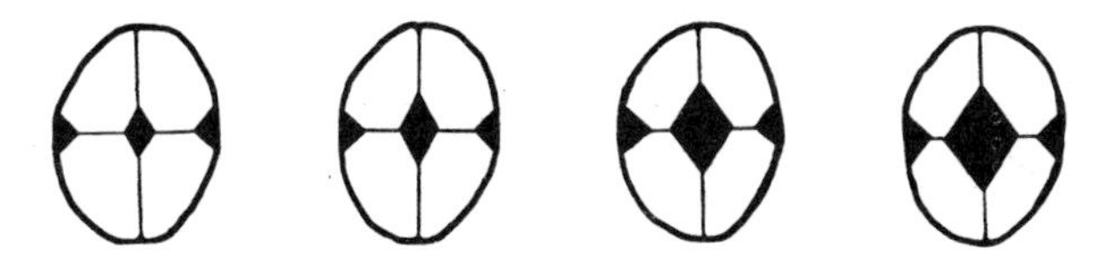

Chiastolite cross sections

Chiastolite. A variety of the mineral andalusite, containing dark, carbonaceous inclusions arranged in regular, cross-shaped design. Good crystals are found mainly in South Australia and Massachusetts.

Childrenite series. A series of orthorhombic, phosphate minerals, which probably grade completely into each other. It includes the following members:

Childrenite	$(Fe,Mn)Al(PO_4)(OH)_2 \cdot H_2O$
Eosphorite	$(Mn,Fe)Al(PO_4)(OH)_2 \cdot H_2O$

Chile saltpeter. A popular name for soda niter, referring to its occurrence in Chile.

Chinese cat's-eye. The shell of a marine snail, showing an eye pattern.

Chloanthite. An isometric mineral, a member of the skutterudite series, composed of nickel-cobalt arsenide, $(Ni,Co)As_3$. It has a silver-gray color.

Chlorapatite. A variety of the mineral apatite, $Ca_5Cl(PO_4)_3$.

Chlorastrolite. A green gem stone. It is an opaque mixture of silicate minerals, the chief of which is pumpellyite. The source is Michigan.

Chlorite group. A group of silicate minerals that crystallize in the monoclinic system and have the same atomic structure and similar physical properties, especially flexibility. The following members are the best known:

Chlorite	$Mg_3(Si_4O_{10})(OH)_2 \cdot Mg_3(OH)_6$
Clinochlore	$(Mg,Al)_3(AlSi_3O_{10})(OH)_2 \cdot Mg_3(OH)_6$
Penninite	$(Mg,Fe,Al)_6(Si,Al)_4O_{10}(OH)_8$

Chloritoid. A member of the brittle-mica group of minerals. It is dark green and is found in metamorphic rocks. The formula is $(Fe,Mg)_2Al_4Si_2O_{10}(OH)_4$, and the crystallization is monoclinic.

Chloromelanite. A dark-green variety of the jade mineral jadeite. Its iron content replaces some of the aluminum.

Chlorophane. A variety of fluorite that fluoresces when heated.

Chlorophoenicite group. A group of arsenate minerals that crystallize in the monoclinic system and have the same atomic structure and a similar occurrence. It includes the following members:

Chlorophoenicite	$(Mn,Zn)_5(AsO_4)(OH)_7$
Magnesium-chlorophoenicite	$(Mg,Mn)_5(AsO_4)(OH)_7$

Chlorospinel. A green variety of spinel.

Chlor-sapphire. The green, gem variety of corundum.

Chondrodite. A monoclinic, silicate mineral occurring mostly in separate grains in marble. It is a member of the chondrodite group. Its color is pale yellow to red. Italy, Sweden, Finland, and New York are good localities, where it is associated with graphite, phlogopite mica, spinel, and pyrrhotite. The formula is $Mg_5(SiO_4)_2(F,OH)_2$.

Chondrodite group. A group of silicate minerals that crystallize in the orthorhombic and monoclinic systems and have closely related atomic structures and similar physical properties. It includes the following members:

Chondrodite	$Mg_5(SiO_4)_2(F,OH)_2$
Norbergite	$Mg_3(SiO_4)(F,OH)_2$
Humite	$Mg_7(SiO_4)_3(F,OH)_2$
Clinohumite	$Mg_9(SiO_4)_4(F,OH)_2$

Chromite. An isometric mineral composed of iron-chromium oxide, $(Fe)Cr_2O_4$. Although chromite occurs in black, often pitchy-looking masses, it has a brown streak. It is the only ore of chromium and is used for bricks in metallurgical furnaces. Turkey, South Africa, the Soviet Union, the Philippines, Southern Rhodesia, Yugoslavia, and New Caledonia share nearly all the world's production.

Chromite series. A series of oxide minerals that crystallize in the isometric system and have the same atomic structure and related physical properties, according to the chemical composition. It belongs to the spinel group and includes the following members:

Chromite	$MgCr_2O_4$
Magnesiochromite	$FeCr_2O_4$

Chrysoberyl. An outstanding gem stone, owing to its hardness (8½), transparency, and interesting optical properties. The colors are in the range of green, brown, and yellow. Alexandrite is green but looks red in artificial light. Cat's-eye and cymophane show chatoyancy. The main sources of chrysoberyl are in Ceylon, the Soviet Union, and Brazil. It is an orthorhombic mineral composed of beryllium-aluminum oxide, $BeAl_2O_4$.

Chrysocolla. An attractive gem mineral. Its blue and green colors and the vitreous or waxy luster resemble those of turquoise; the hardness (2 to 4) is less. The composition is hydrous copper silicate, $CuSiO_3 \cdot 2H_2O$. Compact masses of chrysocolla occur in copper deposits, as in Arizona and New Mexico, where it is a minor ore of copper.

Chrysolite. An older synonym for the mineral olivine. The name is used especially when not referring to olivine as a rock-forming mineral.

Chrysoprase. Green chalcedony quartz. It is colored by nickel oxide.

Chrysotile. The fibrous variety of serpentine. It constitutes the most important kind of asbestos. Large amounts are mined in Quebec, South Africa, the Soviet Union, California, and Arizona.

Cinnabar. The red color and heaviness are distinctive, but cinnabar may be present only as deceptively low-grade crusts or coatings. It is a hexagonal mineral composed of mercury sulfide, HgS. Hepatic cinnabar can be set afire. Cinnabar is the only important ore of mercury. Spain (Almaden) has the world's largest deposit. All localities are associated with recent volcanic activity or hot springs.

Cinnamon stone. A gem variety of grossularite garnet. The brown color of cinnamon is typical. Ceylon is the main locality.

Citrine. Yellow quartz. It is colored by iron and is the gem usually sold as "topaz," in which case it is often obtained by heating an inferior quartz in order to "color it up." Most of it comes from Brazil and Madagascar.

Clausthalite. A mineral composed of lead selenide, PbSe. It occurs in heavy, metallic-gray masses and has come from Germany, Spain, Sweden, and Argentina.

Clay family. A family of hydrous aluminum silicate minerals having somewhat related atomic structures and chemical compositions, as well as similar physical properties, especially plasticity when wet. The most common clay minerals belong to the following groups:

Kaolin group
Montmorillonite group
Attapulgite group
Illite group

Clay iron ore, Clay iron-stone. Miners' names for siderite (an iron ore) mixed with clay.

Cleavelandite. A platy variety of albite (plagioclase) feldspar.

Clinochlore. A member of the chlorite group of minerals. The formula is $(Mg,Al)_3(AlSi_3O_{10})(OH)_2 \cdot Mg_3(OH)_6$.

Clinozoisite. A member of the epidote-clinozoisite series of minerals in the epidote group. It is gray, green, or pink and has a vitreous luster. The formula is $Ca_2Al_3O(SiO_4)(Si_2O_7)(OH)$.

Cobalt bloom. An old name for erythrite, because of its metal content and occurrence as surface growths.

Cobaltite. An isometric mineral composed of cobalt sulfarsenide, CoAsS. It is a member of the cobaltite group. Ontario (Cobalt) and Sweden are large sources, but most of the cobalt of commerce is obtained as a byproduct of copper mining in the Congo.

Cobaltite group. A group of sulfarsenide and similar minerals that crystallize in the isometric system and have the same atomic structure and physical properties. It includes the following members:

Cobaltite	(Co,Fe)AsS
Gersdorffite	NiAsS
Ullmannite	NiSbS

Cobalt vitriol. An old name for bieberite, referring to its metal content and sulfate composition.

Cockscomb pyrites. An old name for marcasite having a cockscomb habit, alluding to its resemblance to pyrite.

Coesite. A silicia mineral, SiO_2, formed by meteorite impact. It is known only in microscopic specimens.

Cog-wheel ore. A British miners' name for bournonite, referring to the shape of its twinned crystals.

Colemanite. A monoclinic, borate mineral. Vitreous, colorless crystals and colorless or white masses occur, usually with ulexite. California and Nevada contain the known localities. The formula is $Ca_2B_6O_{11}{\cdot}5H_2O$.

Collophane. Massive apatite occurring in fossil bones and phosphate rock. It occurs in large deposits in northern Africa, Idaho, and Florida.

Colorado. A name added to several gems to suggest a more valuable gem. Examples include *Colorado jade* for the amazonstone variety of microcline feldspar, and *Colorado ruby* for pyrope garnet.

Columbite. An orthorhombic mineral composed of an oxide of columbium, tantalum, ferrous iron, and manganese, $(Fe,Mn)(Cb,Ta)_2O_6$. It is found in iron-black crystals. It forms a complete series with tantalite and is an ore of both of its rare metals, columbium and tantalum. Columbite comes from pegmatites and placers in the United States, Norway, Madagascar, Western Australia, Japan, and other countries.

Columbite-tantalite series. A series of oxide minerals that crystallize in the orthorhombic system and have the same crystal system and similar physical properties, according to the chemical composition, which grades completely from one end of the series to the other. It includes the following members:

Columbite	$(Fe,Mn)(Cb,Ta)_2O_6$
Tantalite	$(Fe,Mn)Ta_2O_6$

Common mica. Muscovite, so called from its abundance.

Compact black manganese ore. An old name for psilomelane and other manganese minerals, because of their appearance and use.

Connellite group. A group of sulfate minerals that crystallize in the hexagonal system and have the same atomic structure and similar physical properties. It includes the following members:

Connellite	$Cu_{19}(SO_4)Cl_4(OH)_{32}{\cdot}3H_2O$
Buttgenbachite	$Cu_{19}(NO_3)_2Cl_4(OH)_{32}{\cdot}3H_2O$

Copiapite group. A group of sulfate minerals that crystallize in the triclinic system and have the same atomic structure. They grade freely into one another to make several natural series. The members are as follows:

Copiapite	$(Fe,Mg)Fe_4(SO_4)_6(OH)_2 \cdot 20H_2O$
Magnesiocopiapite	$(Mg,Fe)Fe_4(SO_4)_6(OH)_2 \cdot 20H_2O$
Cuprocopiapite	$CuFe_4(SO_4)_6(OH)_2 \cdot 20H_2O$

Copper arsenide group. A group of arsenide minerals that crystallize in different systems and have different properties but resemble each other and occur together. It includes the following members:

Algodonite	Cu_6As
Domeykite	Cu_3As

Copper glance. An old name for chalcocite, referring to its chemical composition and bright, metallic luster.

Copper mica. An old name for chalcophyllite, because of its metal content and micaceous cleavage.

Copper nickel. A miners' name for niccolite, from its coppery color and nickel content.

Copper pitch ore. An old name for tenorite, from its chemical composition and pitchy luster.

Copper pyrites. A familiar name for chalcopyrite, because it contains copper but resembles pyrite.

Cordierite. A blue or violet, vitreous mineral occasionally used as a gem. It is a silicate of magnesium and aluminum, $Mg_2Al_3(AlSi_5O_{18})$, occurring usually in irregular grains or masses. It is hard (7 to 7½) and closely resembles quartz. The gem material comes from Ceylon. The localities for other specimens include Germany, Finland, Greenland, and Madagascar. Iolite and dichroite are other names for cordierite.

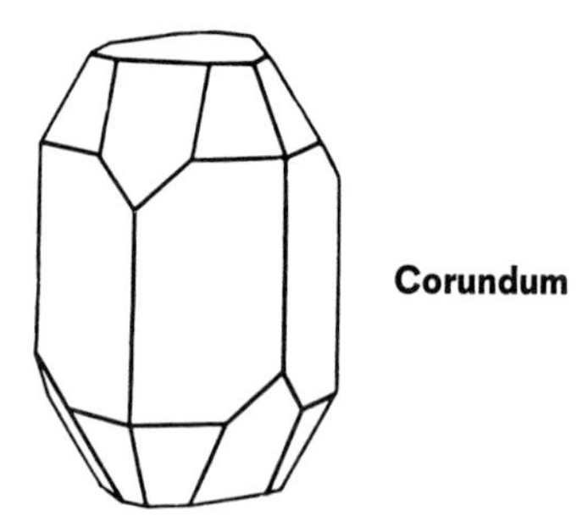
Corundum

Corundum. One of the major gem minerals, having as its varieties ruby and sapphire and the star varieties of each. It is a hexagonal mineral and is recognized by its barrel-shaped crystals, range of colors, and extreme hardness (9), which is next to diamond, making it useful as an abrasive when mixed with magnetite in the form of emery. Corundum comes in large crystals from the Transvaal (South Africa). Large masses occur in North Carolina and Ontario. The gem material comes mainly from the Orient. The formula is Al_2O_3.

Cotton-ball borax. Ulexite, so called because of its appearance and chemical composition.

Covellite. A hexagonal mineral composed of copper sulfide, CuS. Although metallic blue in color, it has a black streak. It occurs in platy masses with other copper minerals. Fine crystals have come from Montana (Butte) and Sardinia.

Covellite group. A group of sulfide-selenide minerals that crystallize in the hexagonal system and are similar in physical properties, although klockmannite is little known. The following members are included:

Covellite	CuS
Klockmannite	CuSe

Cristobalite. A mineral of the silica group, SiO_2. It occurs as rounded aggregates in cavities in volcanic rock. Colorado is a leading locality.

Crocidolite. The blue, asbestos variety of riebeckite. It is mined in South-West Africa and Western Australia. When oxidized and replaced by silica, it becomes the gem called tiger's-eye.

Crocoite. A beautiful chromate mineral, $PbCrO_4$, occurring in bright, orange crystals. These show striations and are noticeably heavy. The one famous locality is Dundas, Tasmania (Australia).

Cross stone. A popular name for twinned crystals of staurolite. Also, for the chiastolite variety of andalusite.

Cryolite. A monoclinic mineral composed of sodium-aluminum fluoride, Na_3AlF_6. Usually the color is white and the luster greasy. It is used in the aluminum industry and as an insecticide. Cryolite comes from Greenland (Ivigtut) and the Soviet Union (Miask).

Cube ore. An old name for pharmacosiderite (a hydrated arsenate of iron), because of its crystal form and occasional use.

Cubic zeolite. An old name for chabazite and analcime, describing the crystal form of these zeolite minerals.

Cummingtonite. A member of the cummingtonite series of minerals in the amphibole family. It is brown and vitreous to silky. It is a hydrous silicate of iron and magnesium, $(Fe,Mg)_7Si_8O_{22}(OH)_2$.

Cummingtonite series. A series of monoclinic, silicate minerals in the amphibole family. It includes the following members:

Cummingtonite	$(Fe,Mg)_7Si_8O_{22}(OH)_2$
Grunerite	$Fe_7Si_8O_{22}(OH)_2$

Cupid's darts. A popular name for rock-crystal quartz enclosing needlelike crystals of other minerals. A more prosaic name is *sagenite.*

Cupreous bismuth. An old name for wittichenite and aikinite, referring to their chemical composition.

Cupreous manganese. An old name for copper-bearing wad, referring to its chemical composition.

Cuprite. An isometric mineral composed of copper oxide, Cu_2O. Its red crystals and masses are heavy. The chalcotrichite variety occurs in brilliant fibers known as plush copper. This ore of copper comes from mines in Arizona, Nevada, Chile, New South Wales (Australia), and the Congo. It is often closely associated with native copper.

Cymophane. A cat's-eye variety of chrysoberyl.

Cyprine. The blue variety of idocrase.

Danburite. A gem mineral, a silicate of calcium and boron, $Ca(B_2Si_2O_8)$. It is hard and occurs in orthorhombic, vitreous crystals that are colorless or pale yellow. Switzerland, Japan, and Madagascar are the leading sources.

Dark red silver ore. A miners' name for pyrargyrite, indicating its color (in comparison with proustite) and use.

Datolite. A boron silicate mineral occurring in two unlike ways: either as monoclinic, vitreous crystals having a pale-green color; or as white masses that look like porcelain. Datolite is found in lava rock, associated with other secondary minerals, such as zeolites, calcite, prehnite, and apophyllite. The countries that yield good datolite include the United States, Norway, Italy, and Germany. The formula is $CaB(SiO_4)(OH)$.

Demantoid. The green, gem variety of andradite garnet. This mineral occurs in the Ural Mountains and is sometimes called Uralian emerald.

Derbyshire spar. A popular name for fluorite from Derbyshire, England, referring to its bright luster on the cleavage surface.

Descloizite group. A group of vanadate minerals that crystallize in the orthorhombic system and have the same atomic structure.

It includes the following members, to which may perhaps be added calciovolborthite and turanite:

Descolizite series	
Pyrobelonite	$MnPb(VO_4)(OH)$

Descloizite series. A series of vanadate minerals in the descloizite group, which grade completely into each other and have a similar origin and occurrence. It includes the following members:

Descloizite	$(Zn,Cu)Pb(VO_4)(OH)$
Mottramite	$(Cu,Zn)Pb(VO_4)(OH)$

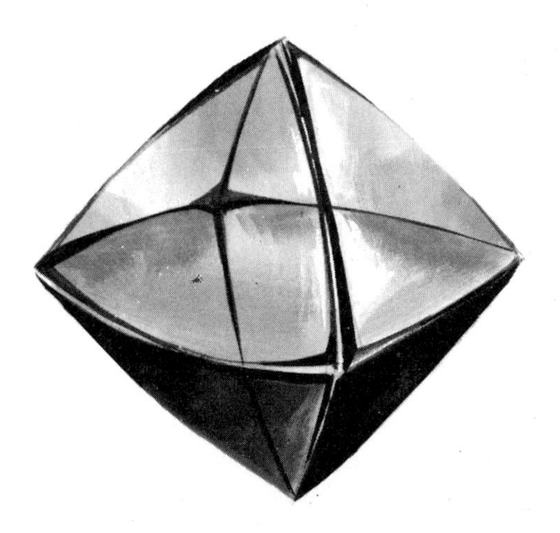

Diamond crystal

Diamond. The hardest known substance. Its crystals (isometric) are often surprisingly large; the biggest was the Cullinan of 3,106 carats (about 22 ounces), which was cut into nine large and ninety-six smaller stones. After being cleaved and sawed to the desired size, diamonds are cut into gem stones, usually in the brilliant form. Diamond is also an important industrial mineral, used in various ways as an abrasive, in wire drawings, and in tools for truing grinding wheels. The industrial varieties of diamond are called *bort* and *carbonado*. India, the original major source of diamond, was succeeded by Brazil, which in turn was replaced by South Africa. The Congo is the main supplier of industrial stones. Angola, Ghana, and a few countries outside Africa yield diamonds, and there is an inactive mine in Arkansas. Diamonds have recently been synthesized—but so far only in small sizes.

Three stages in the cutting of a diamond: First, marking the stone for cutting, a skill that requires firsthand knowledge of diamond crystallography and of the particular stone to be cut. Second, sawing the stone in two with an industrial-diamond saw blade. Owing to diamond's extreme hardness, this process takes several hours. Third, faceting the stone—the diamond is held at precise angles against a rapidly revolving, diamond-dust-impregnated wheel. This forms the 58 facets of the usual brilliant cut.

Diaspore. An orthorhombic mineral composed of hydrogen-aluminum oxide, $HAlO_2$. Its thin crystals are hard and white. Diaspore occurs with corundum and in bauxite, usually in a massive variety not readily recognized by itself. Arkansas, Missouri, France, and Hungary have large-scale deposits.

Dichroite. Another name for cordierite, indicating its property of dichroism.

Dickite. A clay mineral of the *kaolin group*. It usually occurs in veins. The formula is $Al_2Si_2O_5(OH)_4$.

Digenite. An isometric mineral, metallic blue in color, composed of copper sulfide, Cu_9S_5. It occurs in masses, especially in Alaska (Kennecott).

Diopside. A member of the pyroxene family of minerals. It is a silicate of calcium and magnesium, $CaMg(Si_2O_6)$. The luster is vitreous and the color white or light green (due to iron). As the iron content increases, diopside grades into hedenbergite; augite, with aluminum present, is an intermediate member of the diopside-hedenbergite series. Diopside is a metamorphic mineral, found in excellent monoclinic crystals in Switzerland, Austria, Italy, the Ural Mountains, Connecticut, and New York.

Diopside-hedenbergite series. A series of monoclinic, silicate minerals in the pyroxene family, which grade completely into each other. It includes the following members:

Diopside	$CaMg(Si_2O_6)$
Augite	$Ca(Mg,Fe,Al)(Al,Si)_2O_6$
Hedenbergite	$CaFe(Si_2O_6)$

Dioptase. A copper mineral occurring in rich-green crystals. The vitreous rhombohedrons are among the most handsome of specimens. The composition is hydrous copper silicate, $Cu_6(Si_6O_{18}) \cdot 6H_2O$. The best known locality is South-West Africa.

Disthene. Another name for the mineral kyanite, because of its variable hardness, which is what the name means.

Dogtooth spar. A familiar name for pointed crystals of calcite.

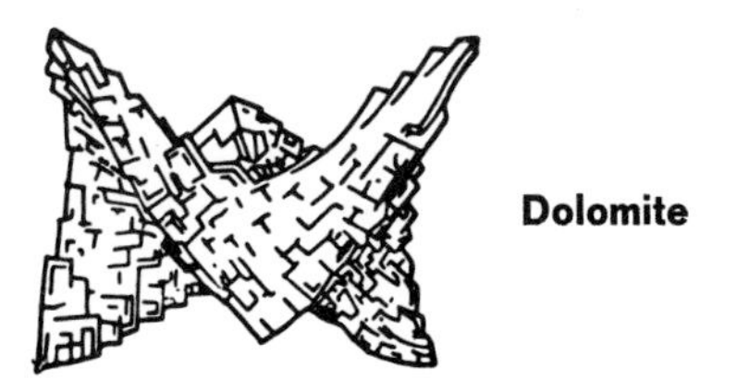

Dolomite

Dolomite. A mineral of the dolomite group, $CaMg(CO_3)_2$. It is typically pink and vitreous or pearly, but it may occur in other hues. The crystals are often curved, being saddle-shaped, or else they grow in rhombs. Austria, Switzerland, Italy, England, and Mexico are major sources.

Dolomite group. A group of carbonate minerals that crystallize in the hexagonal system and have the same atomic structure and similar physical properties. It includes the following members, to which might be added kutnahorite:

Dolomite	$CaMg(CO_3)_2$
Ankerite	$CaFe(CO_3)_2$

Doubly refracting spar. Another name for Iceland spar, because of its distinctive optical property and its bright luster on the cleavage surface.

Dravite. The brown variety of tourmaline.

Dry-bone. An American miners' name for smithsonite, describing its odd appearance.

Dumortierite. A massive mineral, often fibrous. It is a silicate of boron and aluminum, $(Al,Fe)_7O_3(BO_3)(SiO_4)_3$. It is rather hard

(7), and vitreous. Blue, violet, and pink are its colors. Nevada is a source of dumortierite mined as a refractory; California, Mexico, Brazil, Madagascar, and France are other localities.

Dyscrasite. An orthorhombic mineral composed of silver antimonide, Ag_3Sb. Its metallic-white color usually tarnishes. Heavy masses occur in Germany, France, New South Wales (Australia), Nevada, and Ontario.

Earthy cobalt. An old name for cobalt-bearing wad, referring to its appearance and chemical composition.

Earthy ocher of manganese. An old name for wad, indicating its appearance and metal content.

Efflorescing zeolite. An old name for laumontite, because of the flowering appearance of this zeolite mineral.

Egyptian jasper. Jasper having yellow or brown stripes.

Ekanite. A recently discovered gem mineral. It is green, translucent, and radioactive. The composition is calcium-thorium silicate. The source is Ceylon.

Electrum. A natural alloy of gold with more than 20 percent silver. This mineral is pale yellow in color.

Eleolite. The massive and coarsely crystalline variety of nepheline.

Emerald. The green, gem variety of the mineral beryl. Chromium oxide is the coloring matter that gives the magnificent hue. Colombia, where emerald occurs in limestone, has long been the principal locality; Siberia, where it is found in schist, ranks second.

Emerald nickel. An old name for zaratite, because of its color and metal content.

Emery. A hard mixture of minerals, used as an abrasive. Typical emery contains corundum, magnetite (or hematite), and spinel; the blackish magnetite can be separated by passing a magnet over the powder, which then loses its black color. Greece and Turkey are the leading sources of emery.

Enargite. An orthorhombic mineral composed of copper-arsenic sulfide, Cu_3AsS_4. It is a member of the enargite group. Its metallic-gray crystals and masses show a good cleavage. Enargite is an ore of copper, especially in Montana (Butte), and a source of arsenic oxide.

Enargite group. A group of related sulfosalt minerals. Enargite crystallizes in the orthorhombic system, but the system of famatinite is uncertain. The following members are included:

Famatinite	Cu_3SbS_4
Enargite	Cu_3AsS_4

Endlichite. A yellow, arsenic-bearing variety of vanadinite. New Mexico is a noted source.

Enstatite. A member of the enstatite-hypersthene series of minerals in the pyroxene family. It is a magnesium silicate $Mg_2(Si_2O_6)$. It is vitreous and white or pale green. Enstatite occurs in meteorites, as well as in many basic rocks in the earth. New York, Maryland, Pennsylvania, and North Carolina are good localities.

Enstatite-hypersthene series. A series of orthorhombic, silicate minerals in the pyroxene family, which grade completely into one another. It includes the following members:

Enstatite	$Mg_2(Si_2O_6)$
Bronzite	$(Mg,Fe)_2(Si_2O_6)$
Hypersthene	$(Fe,Mg)_2(Si_2O_6)$

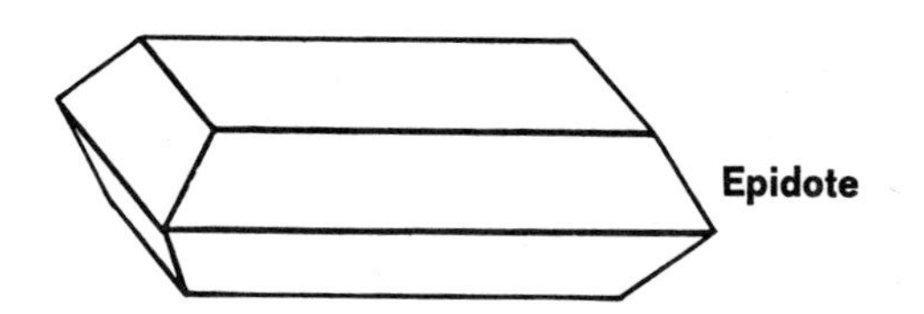

Epidote. An interesting mineral with a peculiar green color. The typical shade is like pistachio, but it is not always the same. The vitreous crystals, which shows striations, are sometimes very handsome, and occur mainly in metamorphic rocks. Notable localities are in Alaska, Austria, Italy, and France. The composition is hydrous silicate of calcium, aluminum, and iron, $Ca_2(Al,Fe)Al_2O(SiO_4)(Si_2O_7)(OH)$. As the iron content decreases, epidote grades into clinozoisite, constituting a series.

Epidote-clinozoisite series. A series of monoclinic, silicate minerals in the epidote group which grade into each other. The members are:

Epidote	$Ca_2(Al,Fe)Al_2O(SiO_4)(Si_2O_7)(OH)$
Clinozoisite	$Ca_2Al_3O(SiO_4)(Si_2O_7)(OH)$

Epidote group. A group of silicate minerals that crystallize in the monoclinic system and have the same atomic structure. It includes the following members, as well as the orthorhombic mineral zoisite:

Epidote-clinozoisite series
Piedmontite $Ca_2(Al,Fe,Mn)Al_2O(SiO_4)(Si_2O_7)(OH)$
Allanite $(Ca,Ce,La,Na)_2(Al,Fe,Mn,Be,Mg)_3O(SiO_4)Si_2O_7)(OH)$

Epsomite. Natural epsom salt in mineral form. The delicately fibrous crusts are colorless or white and vitreous or earthy. The bitter taste is strong. It deposits by evaporation of water in lakes and in caves and mines. British Columbia, Washington, California, Nevada, Wyoming, and Spain have important deposits. The crystallization is orthorhombic, and the formula is $MgSO_4{\cdot}7H_2O$.

Epsomite group. A group of sulfate minerals that crystallize in the orthorhombic system and have the same atomic structure and a similar origin and occurrence. They grade into one another more or less completely to make several natural series. The members are as follows, to which might be added tauriscite:

Epsomite $MgSO_4{\cdot}7H_2O$
Goslarite $ZnSO_4{\cdot}7H_2O$
Morenosite $NiSO_4{\cdot}7H_2O$

Epsom salt. The popular name for epsomite, because it is used in medicine.

Erythrite. A member of the erythrite-annabergite series of minerals in the vivianite group. It occurs as red to pink crusts. The principal localities are in Germany, Ontario, and Idaho. The crystallization is monoclinic, and the formula is $(Co,Ni)_3(AsO_4)_2{\cdot}8H_2O$.

Erythrite-annabergite series. A series of minerals in the vivianite group, which grade completely into each other and have a similar origin. It includes the following members:

Erythrite $Co_3(AsO_4)_2{\cdot}8H_2O$
Annabergite $Ni_3(AsO_4)_2{\cdot}8H_2O$

Erythrosiderite series. A series of orthorhombic, halide minerals which grade completely into each other and have a similar origin and occurrence. It includes the following members:

Erythrosiderite $K_2FeCl_5{\cdot}H_2O$
Kremersite $(NH_4K)_2FeCl_5{\cdot}H_2O$

Eschynite-priorite series. A series of oxide minerals that crystallize in the orthorhombic system and have the same atomic structure and similar physical properties, according to the chemical composition. It includes the following members:

Eschynite	$(Ce,Ca,Fe,Th)(Ti,Cb)_2O_6$
Priorite	$(Y,Er,Ca,Fe,Th)(Ti,Cb)_2O_6$

Euclase. A rare, beryllium silicate mineral, $BeAl(SiO_4)(OH)$. It occurs in prismatic, monoclinic crystals colored yellow, green, or blue.

Euxenite-polycrase series. A series of oxide minerals that crystallize in the orthorhombic system and have the same atomic structure and similar physical properties, according to the chemical composition. It includes the following members:

Euxenite	$(Y,Ca,Ce,U,Th)(Cb,Ta,Ti)_2O_6$
Polycrase	$(Y,Ca,Ce,U,Th)(Ti,Cb,Ta)_2O_6$

Evening emerald. A wrong name for peridot, alluding to its color.

Eye agate. Agate showing a concentric pattern resembling a target.

Eyestone. A popular name for thomsonite.

Fairfieldite group. A group of phosphate minerals that crystallize in the triclinic system and have the same atomic structure. It includes the following members:

Fairfieldite	$Ca_2(Mn,Fe)(PO_4)_2 \cdot 2H_2O$
Collinsite	$Ca_2(Mg,Fe)(PO_4)_2 \cdot 2H_2O$

Fairy stone. A local name for twinned crystals of staurolite.

False. A name added to several gems to indicate their resemblance to a more valuable gem. Examples include *false amethyst, false emerald,* and *false topaz* for fluorite of similar colors, and *false lapis* for lazulite or dyed jasper.

False galena. An English and American miners' name for sphalerite, because of its frequent resemblance to galena.

Famatinite. A mineral composed of copper-antimony sulfide, Cu_3SbS_4. This metallic-gray mineral, a member of the enargite group, occurs with enargite in various localities around the world.

Fayalite. The iron member of the olivine series of minerals, in the olivine group. It occurs as brown to black grains and is less common than forsterite.

Feather ore. A miners' name for several fibrous minerals having a feathery look. Jamesonite, boulangerite, and capillary stibnite are the best known, and zinkenite and meneghinite are also included. Jamesonite and stibnite are *brittle feather ore;* the rest are *flexible feather ore.*

Feather zeolite. An old name for mesotype, thomsonite, and mordenite, because of the occasional feathery appearance of these zeolite minerals.

Feldspar group. The most important and abundant of all mineral groups. Feldspar occurs in almost every major kind of rock; the igneous rocks are named according to the particular feldspar that they contain. The crystals—either monoclinic or triclinic—are often twinned, in a manner of much significance to the petrographer. Feldspar is used in ceramics and glass and as gems. The feldspars are classified as follows:

Potassium feldspars	
Orthoclase	$KAlSi_3O_8$
Sanidine	$(K,Na)AlSi_3O_8$
Microline	$KAlSi_3O_8$
Plagioclase series	
Alkali feldspars	
Potassium feldspar	
Albite	

Feldspathoid group. A group of silicate minerals having somewhat related atomic structures and chemical compositions. They resemble the feldspars but form in rocks of lower silica content. The following members are included, as well as others less familiar:

Leucite	$K(AlSi_2O_6)$
Nepheline	$(Na,K)(AlSiO_4)$
Sodalite	$Na_4(AlSiO_4)_3Cl$
Cancrinite	$Na_8(AlSiO_4)_6(HCO_3)_2$
Lazurite	$(Na,Ca)_4(AlSiO_4)_3(SO_4,S,Cl)$
Hauynite	$(Na,Ca)_{6-8}Al_6Si_6O_{24}(SO_4)_{1-2}$
Noselite	$Na_4Al_3Si_3O_{12} \cdot SO_4$
Petalite	$Li(AlSi_4O_{10})$

Ferberite. A member of the wolframite series of minerals. It is heavy, black, and submetallic looking. Colorado is the main source of this ore of tungsten. The formula is $FeWO_4$, and the crystallization is monoclinic.

Fergusonite series. A series of oxide minerals that crystallize in the tetragonal system and have the same atomic structure and similar physical properties, according to the chemical composition. It includes the following members:

Fergusonite	$(Y,Er,Ce,Fe)(Cb,Ta,Ti)O_4$
Formanite	$(U,Zr,Th,Ca)(Ta,Cb,Ti)O_4$

Fibrolite. The fibrous variety of sillimanite. Some of it resembles green or brown jade and some has a cat's-eye effect.

Figure stone. A popular name for agalmatolite, because of its use in carved Oriental figures and other ornaments.

Fireblende. An old name for pyrostilpnite, because of its red color and bright but nonmetallic luster.

Fire opal. Precious opal having a red to yellow background. The play of color is usually weak, but the principal color may be quite attractive. Mexico is the leading source.

Flèches d'amour. A popular name for rutilated quartz, meaning "arrows of love."

Flexible feather ore. A miners' name for boulangerite, zinkenite, and meneghinite, because of their yielding quality and appearance. The rest of the feather ores are *brittle feather ores.*

Flint. Dull, gray chalcedony. Its conchoidal fracture and sharp edges made it valuable to early man, who used it for implements. It is still used in making gun flints for flintlock guns. The nodules of flint from the chalk cliffs of England and France are especially well known.

Float-stone. An old name for a porous variety of quartz.

Flos ferri. A variety of aragonite occurring in growths resembling coral. It occurs with iron-ore deposits.

Flour gold. Extremely minute particles of native gold. The Snake River, Idaho, is noted for flour gold, which eludes recovery because the particles are so small. An estimated 5,000 pieces would be worth one cent.

Flower agate. Moss agate showing a flowerlike pattern.

Fluorapatite. A variety of the mineral apatite, $Ca_5F(PO_4)_3$.

Fluorite. An isometric mineral composed of sodium fluoride, CaF_2. The crystals (usually cubes) or masses often show bands of color which may vary widely. Handsome specimens come from England, Switzerland, and Illinois. Fluorite, used in metallurgy and chemistry, is mined in many places under the name fluorspar. Blue john and chlorophane are special varieties.

Fluorspar. A miners' name for fluorite, indicating its bright luster on the cleavage surface. This is also the commercial name for fluorite.

Foliated tellurium. An old name for nagyagite, referring to its laminated habit and chemical composition.

Forsterite. The magnesium member of the olivine series of minerals in the olivine group. It is more common than fayalite, occurring as white grains in marble. It is used to make refractory bricks.

Fortification agate. Agate showing an angular pattern resembling the plan of a fort. It is also called *ruin agate.*

Fossil ore. A popular name for hematite occurring in rounded forms, because of its resemblance to fossils.

Fowlerite. A zinc-bearing variety of rhodonite. It is found in large crystals in New Jersey (Franklin area).

Franklinite. An isometric mineral composed of zinc-iron-manganese oxide, $(Zn,FeMn)(FeMn)_2O_4$. Octahedral crystals and masses of this ore of zinc and manganese are abundant in New Jersey (Franklin area) but rare anywhere else. Its mineralogical association with zincite, willemite, and calcite—the last two of which are brilliantly fluorescent—is perhaps the most distinctive of all mineral assemblages known to collectors.

Freibergite. A silver-rich variety of tetrahedrite.

Frondelite series. A series of orthorhombic, phosphate minerals which probably grade completely into each other and have similar physical properties, as well as a similar origin and occurrence. It includes the following members:

Frondelite	$(Mn,Fe)Fe_4(PO_4)_3(OH)_5$
Rockbridgeite	$(Fe,Mn)Fe_4(PO_4)_3(OH)_5$

Gadolinite. A rare, beryllium-silicate mineral, $YFeBe_2(SiO_4)_2O_2$. It is black or brown and has a vitreous luster. It contains yttrium and other rare-earth elements, for which it is a commercial source.

Gahnite. An isometric mineral composed of zinc-aluminum oxide, $ZnAl_2O_4$. It has been called zinc spinel, because it belongs to the spinel series of the spinel group. Large crystals are found in Germany (Bavaria).

Galaxite. An isometric mineral composed of manganese-iron oxide, $(Mn,Fe)Al_2O_4$. It is a member of the spinel series of the spinel group. It has been found in North Carolina.

Galena. A major commercial mineral composed of lead sulfide, PbS. Its heavy, metallic-gray, isometric crystals are often very bright, and it can be one of the most attractive of minerals. Galena has a pronounced cubic cleavage except in the fine-grained material. It is the most important ore of lead and an important ore of silver (a chemical impurity in galena). Missouri has the world's largest deposits of galena. It is also abundant in Idaho, Utah, British Columbia (Canada), Mexico, New South Wales (Australia).

Galena group. A group of sulfide-selenide-telluride minerals that crystallize in the isometric system and have the same atomic structure. The following members are included, the first three have a metallic luster, and the other two do not:

Galena	PbS
Clausthalite	PeSe
Altaite	PbTe
Alabandite	MnS
Oldhamite	(Ca,Fe)S

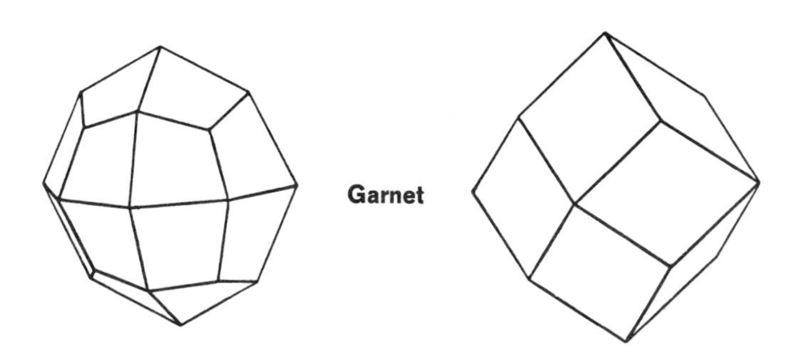
Garnet

Garnet. Any member of an important group of rock-forming and gem minerals. All serve in some capacity as gems, of which rhodolite, demantoid, and cinnamon stone are well-known varieties. Pyrope and almandite are called *precious garnet;* almandite and andradite are called *common garnet.* Melanite is a black variety of andradite. The isometric crystals, substantial hardness (6½ to 7½) and density, and typical range of color from red to yellow help to identify garnet, although the color can be green, white, or black. Garnet is a common abrasive. Besides the original rocks in which it is formed, garnet is also found in placer deposits.

Garnet blende. An old English name for brown sphalerite, from its resemblance to garnet and its bright but nonmetallic luster.

Garnet group. A group of silicate minerals that crystallize in the isometric system and have the same atomic structure and similar physical properties. It includes the following members, which grade more or less into one another:

Pyrope	$Mg_3Al_2(SiO_4)_3$
Almandite	$Fe_3Al_2(SiO_4)_3$
Spessartite	$Mn_3Al_2(SiO_4)_3$
Grossularite	$Ca_3Al_2(SiO_4)_3$
Andradite	$Ca_3Fe_2(SiO_4)_3$
Uvarovite	$Ca_3Cr_2(SiO_4)_3$

Garnierite. A mixture of nickel-bearing silicate minerals. Deposits in New Caledonia and South Africa are an ore of nickel.

Gaylussite. A monoclinic, carbonate mineral, $Na_2Cu(CO_3)_2 \cdot 5H_2O$. It is an evaporite formed by the evaporation of alkali lakes and is found in deserts in the United States and Mongolia.

Gedrite. A member of the anthophyllite series of minerals, in the amphibole family. With a decline in aluminum, it grades into anthophyllite.

Geocronite. An orthorhombic mineral composed of lead-antimony-arsenic sulfide, $Pb_5(Sb,As)_2S_8$. This metallic-gray mineral has been found in Italy, Ireland, and a few other countries.

Gersdorffite. An isometric mineral composed of nickel-arsenic sulfide, NiAsS. It has a metallic-gray color and occurs mostly in crystals. It is a minor ore of nickel in Ontario (Sudbury).

Geyserite. A variety of opal deposited around hot springs and geysers. It is also called siliceous sinter.

Gibbsite. A monoclinic mineral composed of aluminum hydroxide, $Al(OH)_3$. It is an important constituent of bauxite, the ore of aluminum. In this form, it is too fine grained to be identified at sight.

Gibraltar stone. An old name for the Mexican-onyx variety of calcite.

Girasol. Several varieties of opal, moonstone, and other mildly fiery looking gems.

Glance. An old miners' name for minerals having a bright, metallic luster. It has been used in such mineral names as *antimony glance* (stibnite), *silver glance* (argentite), *brittle silver glance* (stephanite), *glance cobalt* (cobaltite).

Glauberite. A monoclinic, sulfate mineral formed by the evaporation of salt lakes. It is vitreous and pale yellow or gray, and it has a somewhat salty taste. Austria, Germany, Chile, and California are among the sources. The formula is $Na_2Ca(SO_4)_2$.

Glauber salt. A popular name for mirabilite, because the artificial salt was discovered by the seventeenth-century chemist, J. R. Glauber.

Glaucodot. An orthorhombic, metallic-gray mineral composed of cobalt-iron sulfarsenide, (Co,Fe)AsS. It is most abundant in Sweden, Norway, and Chile.

Glauconite. A green mineral found in sedimentary rocks of marine origin. It is a complex hydrous silicate related to celadonite and similar in composition to biotite mica, $K_2(Mg,Fe)_2 Al_6(Si_4O_{10})_3(OH)_{12}$.

Glaucophane. A member of the alkali-amphibole or glaucophane-riebeckite series of minerals in the amphibole family. It is often considered one of the soda amphiboles because of its sodium content.

Gmelinite. A member of the zeolite family of minerals, related to chabazite but rarer. The formula is $(Na_2Ca)Al_2Si_4O_{12}\cdot 6H_2O$.

Goethite. An orthorhombic mineral composed of hydrogen-iron oxide, $HFeO_2$. Most of what goes under the name of "limonite" is really goethite—because it is crystalline, as is shown by its fibrous structure or cleavage. The streak is yellowish brown, although the color may range to black. Superb crystals come from Colorado (Pikes Peak region) and Michigan (Negaunee). Massive goethite cannot be separated from limonite on the basis of locality.

Goethite group. A group of oxide minerals that crystallize in the orthorhombic system and have the same atomic structure and related physical properties. It includes the following members:

Diaspore	$HAlO_2$
Goethite	$HFeO_2$

Gold group. A group of native-element minerals that crystallize in the isometric system and have the same atomic structure and similar physical properties. It includes the following members:

Native gold	Au
Maldonite	Au_2Bi
Native silver	Ag

Native copper	Cu
Native lead	Pb

Gold quartz. A variety of milky quartz containing flecks or nuggets of native gold. California has been a common source of this ornamental material.

Goldstone. Artificial glass colored by metallic copper. A similar product is made in other ways.

Grapestone. A popular name for a cluster of chalcedony quartz having a botryoidal habit.

Graphic tellurium. An old name for sylvanite, describing certain twinned crystals of this tellurium mineral that look like writing.

Graphite. Hexagonal, native carbon. Its softness makes it suitable for pencils and lubricants, and its high melting point makes it useful for crucibles. It usually occurs in metallic-gray, sheetlike masses. Some graphite is derived from coal, but most of it is found in veins.

Gray antimony. An old name for stibnite and jamesonite, indicating their color and chemical composition.

Gray cobalt ore. An old name for smaltite, because of its color and use.

Gray copper ore. An old name for tetrahedrite, because of its color and use.

Gray oxide of manganese. An old name for pyrolusite, indicating its color and chemical composition.

Gray sulphuret of copper. An old name for tennantite, because of its color and chemical composition.

Greenalite. A silicate mineral related to serpentine. It occurs in the iron deposits of Minnesota. The formula is $FeSi_4O_{10}(OH)_8$.

Green copper carbonate. Malachite.

Green enargite. A name mistakenly given to tennantite that has altered from enargite.

Green lead ore. An old name for pyromorphite and mimetite, referring to their color and use.

Greenockite. A hexagonal mineral composed of cadmium sulfide, CdS. It is usually found as a powdery, yellow coating on sphalerite, as in the Tri-State district of Missouri, Kansas, and Oklahoma.

Greenstone. A name used for chlorastrolite and nephrite jade, because of their color, but also given to a number of altered igneous rocks—now metamorphic—that are green.

Green vitriol. An old name for melanterite, indicating its color and sulfate composition.

Grossularite. A subspecies of the garnet group of minerals. It is a silicate of calcium and aluminum, $Ca_3Al_2(SiO_4)_3$. The color may be white or green, it also ranges from yellow to red. It occurs mostly in marble; Ceylon is a noteworthy locality. *Cinnamon stone* is a gem variety of grossularite.

Gruenlingite. A soft and heavy mineral occurring in metallic-gray masses. It has been found in England and Spain. It is a member of the tetradymite group. The crystallization is hexagonal, and the formula is Bi_4TeS_3.

Grunerite. A member of the cummingtonite series of minerals in the amphibole family. With an increase in magnesium, it grades into cummingtonite.

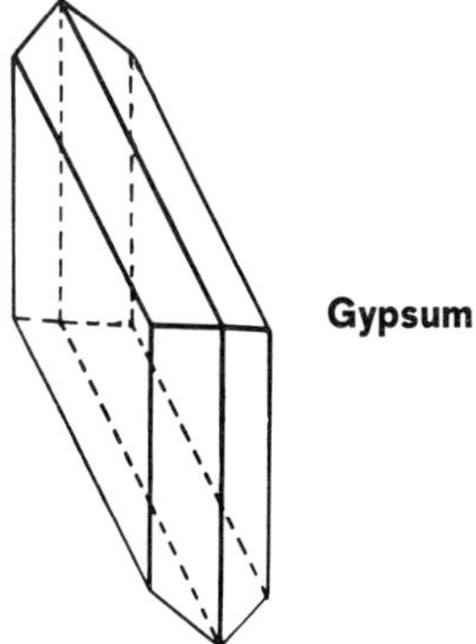

Gypsum

Gypsum. An abundant mineral of worldwide distribution. It is a hydrous calcium sulfate, $CaSO_4 \cdot 2H_2O$, deposited by the evaporation of sea water. It occurs mostly as soft, white masses, although it has several special varieties, including selenite (crystallized), alabaster (compact), and satin spar (fibrous). Gypsum is the basis of the plaster industry.

Hair salt. A popular name for silky, fibrous epsomite.

Hairstone. The sagenite variety of quartz, descriptive of the needlelike crystal inclusions of various minerals.

Halfbreed. Associations of native copper and native silver. These occur in combination in the Keweenaw Peninsula, Michigan. They tarnish alike but can be cleaned to reveal the two separate minerals.

Halite. An isometric mineral composed of sodium chloride, NaCl. It occurs in cubic crystals (sometimes hopper-shaped) and also in granular (often cleavable) or compact masses called rock salt. Its deposits are nearly worldwide, and its uses are extensive in food and in the chemical and other industries.

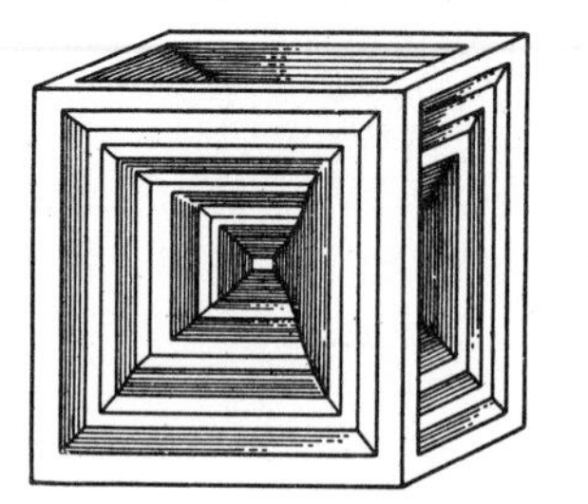

Halite

Halite group. A group of halide minerals that crystallize in the isometric system and have the same atomic structure and similar physical properties. It includes the following members:

Halite	$NaCl$
Sylvite	KCl
Villiaumite	NaF
Cerargyrite series	

Halloysite. A clay mineral of the kaolin group. Its formula is $Al_2Si_2O_5(OH)_4$.

Halotrichite group. A group of sulfate minerals that crystallize in the monoclinic system and have the same atomic structure and similar physical properties. They grade more or less into one another to make several natural series. The members are as follows, to which might be added redingtonite:

Halotrichite	$FeAl_2(SO_4)_4 \cdot 22H_2O$
Pickeringite	$MgAl_2(SO_4)_4 \cdot 22H_2O$
Apjohnite	$MnAl_2(SO_4)_4 \cdot 22H_2O$
Dietrichite	$(Zn,FeMn)Al_2(SO_4)_4 \cdot 22H_2O$
Bilinite	$FeFe_2(SO_4)_4 \cdot 22H_2O$

Harlequin opal. White opal showing even patches of color resembling a mosaic.

Harmotome. A member of the stilbite group of minerals, in the zeolite family. The crystallization is monoclinic, and the formula is $Ba(Al_2Si_6O_{16}) \cdot 6H_2O$.

Hauerite. A mineral of the pyrite group. It is reddish brown but alters to metallic black. Large crystals have come from Sicily. The crystallization is isometric, and the formula is MnS_2.

Hausmannite. A tetragonal mineral composed of manganese oxide, $MnMn_2O_4$. It occurs in brownish-black crystals and granular masses, especially in Germany and Sweden and also in Arkansas. It is an ore of manganese.

Hausmannite group. A group of minerals that crystallize in the tetragonal system and have the same atomic structure and related physical properties. It includes the following members:

Hausmannite	$MnMn_2O_4$
Hetaerolite	$ZnMn_2O_4$

Hauynite. A blue, isometric mineral of the feldspathoid group. The formula is $(Na,Ca)_{6-8}Al_6Si_6O_{24}\cdot(SO_4)_{1-2}$.

Hawk's-eye. A gem variety of quartz. When the original (blue) crocidolite has not been oxidized, as it has in tiger's-eye, this name is used for the blue material.

Heavy sand. Sand containing an appreciable amount of heavy minerals. These include hematite, ilmenite, rutile, magnetite, chromite, monazite, garnet, zircon, and staurolite.

Heavy spar. A popular name for barite, indicating its weight and its bright luster on the cleavage surface.

Hedenbergite. A member of the diopside-hedenbergite series of minerals in the pyroxene family. It is a silicate of calcium and iron. The luster is vitreous and the color is green. As the iron content decreases, hedenbergite grades into diopside. Sweden is a noted source of hedenbergite crystals.

Heliodor. A golden-yellow variety of the mineral beryl. Ceylon and Siberia are good localities.

Heliotrope. The name used in Great Britain for bloodstone.

Hematite (ironrose)

Hematite. The most important iron mineral. It is hexagonal and is composed of iron oxide, Fe_2O_3. It is widely distributed as the coloring matter of red and brown rocks. Huge masses constitute the world's most valuable ore of iron, mined in the Lake Superior region, Quebec, Alabama, Venezuela, and Brazil. Specular hematite, or specularite, is a variety occurring in bright, metallic masses, in flakes like mica, and in crystals. This variety is much used for carving intaglios and for opaque, faceted stones. Regardless of color, hematite powders to a reddish-brown streak.

Hematite group. A group of oxide minerals that crystallize in the hexagonal system, but the ilmenite series has a different crystalline atomic structure. It includes the following members:

Corundum	Al_2O_3
Hematite	Fe_2O_3
Ilmenite series	

Hemimorphite. A zinc mineral occurring in crystal clusters. The opposite ends of a hemimorphite crystal are different, and the groups grow in interesting arrangements. The mineral is vitreous and nearly colorless. Its composition is hydrous zinc silicate, $Zn_4(Si_2O_7)(OH)_2 \cdot H_2O$. New Jersey (Franklin area), Colorado (Leadville), Mexico, Northern Rhodesia, Algeria, and Sardinia are among the localities for outstanding specimens.

Hercynite. An isometric mineral composed of iron-aluminum oxide, $FeAl_2O_4$. It has been called iron spinel, because it belongs to the spinel series of the spinel group.

Herderite series. A series of monoclinic, phosphate minerals, which grade completely into each other and have similar physical properties, as well as a similar origin and occurrence. It includes the following members:

Herderite	$CaBe(PO_4)(F,OH)$
Hydroxyl-herderite	$CaBe(PO_4)(OH,F)$

Herkimer diamond. A wrong name for quartz crystals from Herkimer County, New York. They taper at both ends and are clear and unusually lustrous.

Hessite. A mineral composed of silver telluride, Ag_2Te. Gray crystals and heavy masses have been reported from Colorado, California, Mexico, Chile, Western Australia, Rumania, Asia Minor, and Siberia.

Hessonite. Another name for the cinnamon stone variety of grossularite garnet.

Heterosite series. A series of orthorhombic, phosphate minerals which grade into each other and have similar physical properties, as well as a similar origin and occurrence. It includes the following members:

Heterosite	$(Fe,Mn)PO_4$
Purpurite	$(Mn,Fe)PO_4$

Heulandite. A member of the zeolite family of minerals. It is recognized by the diamond shape of the side face, which has a pearly luster on the cleavage surface. The color is white unless stained red. The crystallization is monoclinic, and the formula is $Ca(Al_2Si_7O_{18}) \cdot 6H_2O$.

Hexahydrite group. A group of sulfate minerals that crystallize in the monoclinic system and have the same atomic structure. It includes the following members:

Hexahydrite	$MgSO_4 \cdot 6H_2O$
Bianchite	$ZnSO_4 \cdot 6H_2O$

Hiddenite. The vitreous, green, gem variety of the mineral spodumene. It is rare, coming almost entirely from North Carolina.

Hieratite group. A group of halide minerals that crystallize in the isometric system and have the same atomic structure and a similar origin and occurrence. They seem also to grade into each other as a complete series. It includes the following members:

Hieratite	K_2SiF_6
Cryptohalite	$(NH_4)_2SiF_6$

Holmquistite. A member of the amphibole group of minerals. Because of its sodium content, it is often known as one of the soda amphiboles.

Hornblende. The most important member of the amphibole family of minerals. It is a hydrous silicate of extremely complex composition. It is vitreous and dark green, dark brown, or black. As the composition and properties range through the hornblende series, special names have been applied; basaltic hornblende contains titanium. Hornblende is widespread throughout the world.

Hornblende series. A series of monoclinic, silicate minerals in the amphibole family. It includes the following members, as well as others less familiar:

Edenite	$NaCa_2Mg_5AlSi_7O_{22}(OH)_2$
Pargasite	$NaCa_2Mg_4Al_3Si_6O_{22}(OH)_2$
Hastingsite	$(Ca,Na,K)_3Fe_5(Si,Al)_8O_{22}(OH)_2$

Horn mercury. An old name for calomel, indicating its appearance and metal content.

Horn silver. A miners' name for cerargyrite, from its resemblance to animal horn.

Horse-flesh ore. An English miners' name for bornite, referring to its distinctive appearance.

Horsfordite. A mineral composed of copper antimonide, Cu_5Sb. It occurs in Greece as heavy, metallic-white masses.

Huebnerite. A member of the wolframite series of minerals. It is heavy, brown, and submetallic to resinous. Nevada, South Dakota, and Colorado are the main sources. The formula is $MnWO_4$, and the crystallization is monoclinic.

Hühnerkobelite-varulite series. A series of phosphate minerals in the triphylite group which apparently grade completely into each other and have similar physical properties, as well as a similar origin and occurrence. It includes the following members:

Hühnerkobelite	$(Na,Ca)(Fe,Mn)_2(PO_4)_2$
Varulite	$(Na,Ca)(Mn,Fe)_2(PO_4)_2$

Hyacinth. A gem variety of the mineral zircon. It is brown or reddish orange, the latter color tending toward jacinth. Hyacinth is also the same as cinnamon stone garnet.

Hyalite. A clear, glassy variety of opal.

Hyalophane. A barium feldspar, $(K,Ba)(Al,Si)_2Si_2O_8$.

Hydrophane. A variety of opal that shows opalescence only when put in water.

Hydrotalcite group. A group of carbonate-hydroxide minerals that crystallize in the hexagonal system and have the same atomic structure (but different from the sjogrenite group) and similar physical properties. It includes the following members:

Hydrotalcite	$Mg_6Al_2(OH)_{16}\cdot CO_3\cdot 4H_2O$
Stichtite	$Mg_6Cr_2(OH)_{16}\cdot CO_3\cdot 4H_2O$
Pyroaurite	$Mg_6Fe_2(OH)_{16}\cdot CO_3\cdot 4H_2O$

Hydroxylapatite. A variety of the mineral apatite, $Ca_5(OH)(PO_4)_3$.

Hypersthene. A member of the enstatite-hypersthene series of minerals, in the pyroxene family. It is a silicate of magnesium and iron, $(Mg,Fe)_2(Si_2O_6)$. The color is brownish green to black, and the luster is vitreous or pearly. Hypersthene grades into bronzite and then into enstatite as the iron content decreases. New York has a number of good localities for hypersthene.

Ice. A hexagonal mineral composed of solid water, H_2O. Its hexagonal symmetry is clearly seen in snow flakes. When occurring in large bodies, ice is a layered sedimentary rock, which is transformed (in glaciers) to a metamorphic rock.

Iceland spar. Transparent, colorless calcite. It is used in optical instruments; the name *doubly refracting spar* indicates its characteristic optical property, as well as its bright luster on the cleavage surfaces. Important sources of it include localities in Iceland and New Mexico.

Ice-stone. A popular name for cryolite, indicating its appearance.

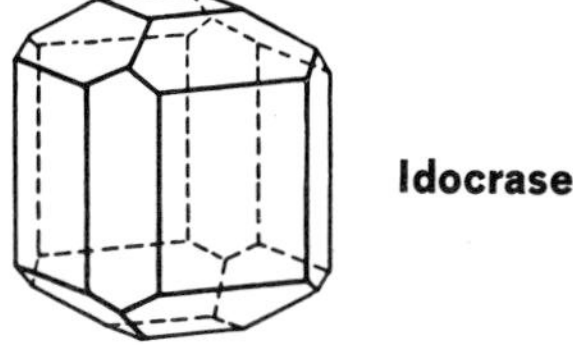

Idocrase

Idocrase. A silicate mineral of metamorphic origin. It occurs mainly in crystalline aggregates, showing striations, and less often in simple tetragonal crystals. The color is usually green or brown. A green, jadelike variety from California is known as californite. Cyprine is a blue variety. Idocrase comes from Mount Vesuvius and elsewhere in Europe, Siberia, Mexico, and the United States. The formula is $Ca_{10}(Mg,Fe)_2Al_4(SiO_4)_5(Si_2O_7)_2(OH)_4$.

Illite. Any clay mineral that resembles mica, although the chemical composition is somewhat different. The structure is often a combination of mica and montmorillonite. Illites are the main constituents in many kinds of shale.

Illite group. A group of clay minerals intermediate between montmorillonite and muscovite mica. Many of them are actual interlayerings of these minerals.

Ilmenite. A hexagonal, titanium-ore mineral composed of iron-titanium oxide, $FeTiO_3$. It has an iron-black color and occurs in masses or individual grains. The United States, Canada, Norway, and India are producers of ilmenite. The world's largest single deposit is in Quebec (Allard Lake).

Ilmenite series. A series of oxide minerals that crystallize in the hexagonal system and have the same atomic structure and similar physical properties. It belongs to the hematite group and includes the following members:

Ilmenite	$FeTiO_3$
Geikelite	$MgTiO_2$
Pyrophanite	$MnTiO_3$

Ilvaite. A mineral related to lawsonite. It is a hydrous silicate of calcium and iron, $CaAl_2(Si_2O_7)(OH)_2 \cdot H_2O$.

Image stone. A popular name for agalmatolite, from its use for carved Oriental figures.

Indicolite. The dark-blue, gem variety of the mineral tourmaline.

Indigo-copper. An old name for covellite, because of its dark-blue color and metal content.

Iodyrite. A hexagonal mineral composed of silver iodide, AgI. A shiny, colorless mineral, it has come in substantial amounts from Chile and New South Wales (Australia).

Iolite. Another name for cordierite, referring to its violet color.

Iridosmine. A natural alloy of iridium and osmium. It is a metallic-white mineral belonging to the iridosmine series and occurring in hard and very heavy flakes and grains. Numerous localities are known in platinum-mining districts around the world.

Iridosmine series. A series of minerals that crystallize in the hexagonal system and have the same atomic structure and similar physical properties, according to the chemical composition, which grades within the middle ranges of the two extremes (end members). When iridium is the chief element, the name is iridosmine; when osmium is in excess, the name is siserskite. The members include the following:

Iridosmine	Ir,Os
Siserskite	Os,Ir

Iris agate. Agate showing rainbow colors from closely spaced, parallel layers which diffract the light.

Iris quartz. Rock crystal containing cracks that make iridescent spots.

Iron. As a terrestrial, metallic-gray mineral, iron occurs in a few localities, especially Greenland (Disko Island). Its most widespread occurrence is in siderite meteorites, in which it is always alloyed with nickel and smaller amounts of cobalt and copper, to make alloys called kamacite and taenite. The iron of commerce is obtained from iron ores: taconite, hematite, magnetite, goethite, limonite, siderite, pyrite.

Iron glance. An old name for specular hematite, indicating its chemical composition and bright, metallic luster.

Iron group. A group of native-element minerals that crystallize in the isometric system and have similar physical properties. It includes the following members:

Native iron	Fe,Ni
Nickel-iron	Ni,Fe

Iron pyrites. A common name for pyrite, emphasizing its iron content.

Iron rose. A variety of hematite occurring in thin plates grouped as a rosette.

Jacobsite. An isometric mineral composed of manganese-iron oxide, $MnFe_2O_4$. Several localities in Sweden produce this dark, metallic-brown mineral.

Jacinth. An orange-colored, gem variety of the mineral zircon. As it becomes brown, it grades into hyacinth.

Jacinth is also used to denote hessonite garnet.

Jade. The outstanding Chinese gem material. True jade may be either jadeite (a pyroxene) or nephrite (an amphibole), but there are a number of other minerals used as jade substitutes. Jade comes in a wide range of colors besides the familiar green. Because of its internal structure, it is the toughest of gems to break. The two main kinds of jade differ in their properties and occurrence, but they are employed for ornamental purposes without distinction.

Jadeite. One of the two minerals constituting true jade. It is a member of the spodumene series in the pyroxene family. It is a silicate of sodium and aluminum, $NaAl(Si_2O_6)$. The green color is characteristic, but white (often with green spots) is also familiar, and brown and violet colors are also known. The vitreous luster is enhanced by polishing the carved pieces. Jadeite is the more valuable of the two jade minerals, the other being nephrite; it is a metamorphic mineral found in masses in Burma, Japan, Guatemala, and California. Chloromelanite is a dark-green variety.

Jamesonite. A monoclinic mineral composed of lead-iron-antimony sulfide, $Pb_4FeSb_6S_{14}$. It has a fibrous look and a metallic-gray color. It is considered an ore of lead at a number of its many localities.

Jargon. A gem variety of the mineral zircon. It has a range of color (smoky, yellow, colorless) and is not too definitely defined. Ceylon is the chief locality.

Jarosite. A brown, sulfate mineral, widespread as crusts and coatings but seldom recognized. The formula is $KFe_3(SO_4)_2(OH)_6$, and the crystallization is hexagonal. Germany and Chile contain considerable amounts.

Jasper. Dark chalcedony. This is the common, opaque form of quartz. It comes in red, brown, yellow, green, and other colors, often in bands or patches. Special varieties go under the names *ribbon jasper* and *Egyptian jasper.* Green jasper is called *prase.*

Jasperized wood. Petrified wood that has been filled or replaced by dark chalcedony of the jasper variety.

Jet. A gem variety of lignite coal. It is black and has a conchoidal fracture. England and Spain are the main sources.

Kainite. A monoclinic mineral composed of hydrous sulfate-chloride, $MgSO_4 \cdot KCl \cdot 3H_2O$. This colorless, glassy mineral has a bitter-salty taste. It is important in Germany as a source of potash.

Kaliophilite, A member of the feldspathoid group of minerals. Its formula is $KAlSiO_4$, and it grades into nepheline.

Kaolin group. A group of clay minerals having the same chemical composition but different atomic structures. It includes the following members:

Kaolinite	$Al_2Si_2O_5(OH)_4$
Dickite	$Al_2Si_2O_5(OH)_4$
Nacrite	$Al_2Si_2O_5(OH)_4$
Halloysite	$Al_2Si_2O_5(OH)_4$

Kaolinite. The best known clay mineral. In association with nacrite and dickite, it makes up the material called kaolin. The formula is $Al_2Si_2O_5(OH)_4$.

Kernite. A monoclinic, borate mineral. It occurs in long, splintery masses which are vitreous or pearly and colorless or white. California (Kern County) is the only source, but the supply of this leading boron mineral is large. The formula is $Na_2B_4O_7 \cdot H_2O$.

Kidney ore. Hematite occurring in reniform habit.

Kieserite. A monoclinic mineral composed of hydrous magnesium sulfate, $MgSO_4 \cdot H_2O$. This white mineral occurs in granular masses. It is found in Germany and the United States (New Mexico, Texas).

Kieserite group. A group of sulfate minerals that crystallize in the monoclinic system and have the same atomic structure and similar physical properties. It includes the following members:

Kieserite	$MgSO_4 \cdot H_2O$
Szomolnokite	$FeSO_4 \cdot H_2O$
Szmikite	$MnSO_4 \cdot H_2O$

Krennerite. An orthorhombic mineral composed of gold-silver telluride, $(Au,Ag)Te_2$. It is heavy and silver colored, but it yields a mixed button of gold and silver when the tellurium is driven off by heat. The localities are Colorado, Quebec, Western Australia, and Rumania.

Krennerite group. A group of telluride minerals that crystallize in two different systems but are similar in their physical properties. It includes the following members:

Krennerite	$AuTe_2$
Calaverite	$AuTe_2$
Sylvanite	$(Ag,Au)Te_2$

Kunzite. The lilac-colored, gem variety of the mineral spodumene. California and Madagascar are the usual localities.

Kyanite. A mineral of considerably varying hardness (5 to 7). It can be scratched with a knife down the cleavage of the long, bladed crystals but not across them. The color is usually in blue and white or green and white streaks, and the luster is vitreous or pearly. An aluminum silicate, $AlAlO(SiO_4)$, kyanite has a metamorphic origin. It is mined as a refractory in India, Kenya, and the United States. Good specimens also occur in Switzerland, Austria, and France.

Labradorite. A member of the plagioclase feldspar series of minerals. It consists of 30 to 50 percent albite and the rest anorthite. Gemmy material shows an iridescent sheen, usually blue or green. It comes from Labrador and elsewhere.

Landscape agate. A less common name for moss agate.

Langbeinite group. A group of sulfate minerals that crystallize in the isometric system and have the same atomic structure. It includes the following members:

Langbeinite	$K_2Mg_2(SO_4)_3$
Manganolangbeinite	$K_2Mn_2(SO_4)_3$

Lapis lazuli. A gem rock composed of several blue minerals of the feldspathoid group, together with pyrite, calcite, and often other minerals, mostly silicates. Excellent material comes from Afghanistan, Siberia, Chile, and Colorado.

Larsenite. A member of the olivine group of minerals. It is an orthorhombic silicate of lead and zinc, $PbZn(SiO_4)$.

Laumontite. A member of the zeolite family of minerals. It is less common than most of the others, except where it occurs as an alteration of feldspar and glass in certain volcanic-ash sediments. The crystallization is monoclinic, and the formula is $(Ca,Na)_7Al_{12}(Al,Si)_2Si_{26}O_{80}\cdot 25H_2O$. Fresh laumontite alters readily to the chalky variety known as leonhardite.

Lawrencite group. A group of halide minerals that crystallize in the hexagonal system and have the same atomic structure and a similar origin and occurrence. It includes the following members:

Lawrencite	$FeCl_2$
Scacchite	$MnCl_2$
Chloromagnesite	$MgCl_2$

Lawsonite. A hard mineral of unusual composition, $CaAl_2(Si_2O_7)(OH)_2\cdot H_2O$. This is like feldspar plus water but it is remarkably hard (8). The white or nearly white crystals and masses are vitreous to greasy. Lawsonite occurs in California, France, Italy, Corsica, and New Caledonia.

Lazulite. A blue, monoclinic mineral of the lazulite series. It occurs in vitreous masses, especially in Austria, Sweden, Madagascar, Georgia, and California. The formula is $(Mg,Fe)Al_2(PO_4)_2(OH)_2$.

Lazulite series. A series of monoclinic, phosphate minerals which probably grade completely into each other and have a similar origin and occurrence. It includes the following members:

Lazulite	$(Mg,Fe)Al_2(PO_4)_2(OH)_2$
Scorzalite	$(Fe,Mg)Al_2(PO_4)_2(OH)_2$

Lazurite. An isometric member of the feldspathoid group of minerals. It is usually found as vitreous, blue masses. The formula is $(Na,Ca)_4(AlSiO_4)_3(SO_4,S,Cl)$. Lapis lazuli is a rock mixture of lazurite and other minerals.

Lead glance. An old name for galena, from its chemical composition and bright, metallic luster.

Lead ocher. An old name for massicot, indicating its chemical composition and earthy look.

Lead vitriol. An old name for anglesite, from its metal and sulfate content.

Lechatelierite. Silica glass formed when lightning fuses sand or rock. The curiously twisted tube that results in loose sand is known as a fulgerite. The same product has been discovered in meteorite craters, where impact was the cause.

Lechosos opal. White opal showing a deep-green play of color.

Lenticular iron-ore. A popular name for hematite occurring in lens-shaped bodies.

Leonhardite. The chalky variety of the zeolite mineral laumontite.

Lepidocrocite. An orthorhombic mineral composed of iron hydroxide, FeO(OH). The small, red crystals usually occur on goethite and are common in Germany.

Lepidocrocite group. A group of oxide minerals that crystallize in the orthorhombic system and have the same atomic structure and similar physical properties. Manganite, MnO(OH), which is monoclinic, may belong because of chemical similarities. The following members are included:

Lepidochrocite	FeO(OH)
Boehmite	AlO(OH)

Lepidolite. The lithium-bearing mica. The usual color is pink to lilac, but this characteristic tint may be absent. It contains fluorine in addition to the fundamental mica formula, which becomes $K_2Li_3Al_3(AlSi_3O_{10})_2{\cdot}(OH,F)_4$. Pegmatite is the typical home. The Soviet Union, Madagascar, Maine, Colorado, and California are among the chief localities.

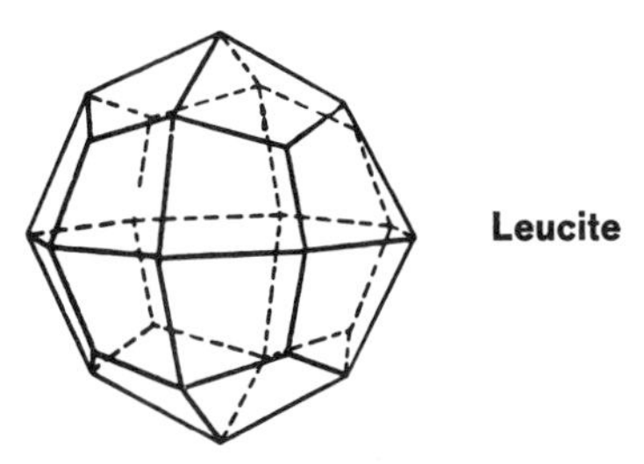
Leucite

Leucite. A member of the feldspathoid group of minerals. Its isometric crystals occur in distinct, embedded grains. They are vitreous or dull, and white or gray. Mount Vesuvius is a noted locality, and Wyoming and Montana are other important localities. The formula is $KAlSi_2O_6$.

Light red silver ore. A miners' name for proustite, indicating its color (in comparison with pyrargyrite) and use.

Limonite. A field name for mixtures of various hydrous iron oxides. Most so-called limonite proves to be really goethite because it is crystalline, as is shown by its fibrous structure or cleavage. The important iron-ore deposits of France (Lorraine) are limonite.

Linnaeite. An isometric mineral composed of cobalt sulfide,

Co_3S_4. The metallic-gray crystals and masses occur in Germany and the Congo.

Linnaeite series. A series of sulfide minerals that crystallize in the isometric system and have the same atomic structure and similar physical properties. It includes the following members:

Linnaeite	Co_3S_4
Siegenite	$(Co,Ni)_3S_4$
Carrollite	Co_2CuS_4
Violarite	Ni_2FeS_4
Polydymite	Ni_3S_4

Litharge. A tetragonal mineral composed of lead oxide, PbO. It occurs as red crusts and has been found in California and Idaho.

Lithia mica. A familiar name for lepidolite, from its chemical composition.

Lithiophilite. An orthorhombic, phosphate mineral of the triphylite-lithiophilite series. It occurs in vitreous to resinous masses that are brown or pink, but often stained nearly black. The chief localities are in pegmatites in Portugal, Sweden, and South Dakota. The formula is $LiMnPO_4$.

Lodestone. A naturally magnetic variety of magnetite. The first compasses were based on the polarity of this mineral, but the cause of the magnetism is still unknown. The best localities for strong lodestone include Germany, Italy (Elba), Transvaal (South Africa), and Siberia.

Loellingite group. A group of arsenide minerals that crystallize in the orthorhombic system, but the status is not certain for any except loellingite. The following members are included:

Loellingite	$FeAs_2$
Safflorite	$(Co,Fe)As_2$
Rammelsbergite	$NiAs_2$
Pararammelsbergite	$NiAs_2$

Ludwigite series. A series of orthorhombic, borate minerals which grade completely into each other and have a similar origin and occurrence. It includes the following members:

Ludwigite	$(Mg,Fl)_2FeBO_5$
Paigeite	$(Fe,Mg)_2FeBO_5$

Magnesioferrite. An isometric mineral composed of magnesium-iron oxide, $MgFe_2O_4$. This metallic-black mineral comes mostly from the volcanic steam vents called fumaroles, as at Vesuvius, Etna, and Stromboli.

Magnesite. A mineral of the calcite group, $MgCO_3$. It is white or gray, yellow or brown. Its use is in bricks for lining metallurgical furnaces and as a source of magnesia. Important localities are in Manchuria, the Soviet Union, Austria, California, Nevada, and Washington.

Magnetic iron ore. A popular name for magnetite, indicating its most distinctive property.

Magnetic pyrites. A miners' name for pyrrhotite, because of its magnetism and pyritic color.

Magnetic sulphuret of iron. An old name for pyrrhotite, because of its magnetism and chemical composition.

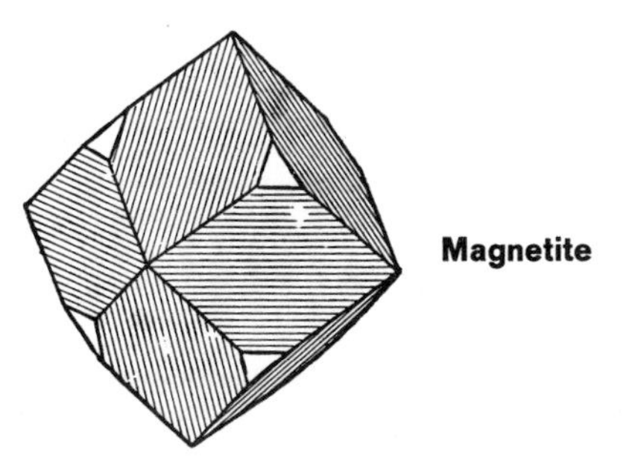

Magnetite

Magnetite. An isometric mineral composed of iron oxide, Fe_3O_4. It belongs to the magnetite series in the spinel group. It is black and strongly magnetic; the lodestone variety is a natural magnet. The most common form of emery is a mixture of magnetite and corundum. When octahedrons of magnetite alter to hematite as pseudomorphs, they are called martite. One of the most abundant oxides, magnetite is a major ore of iron. Huge deposits of magnetite are mined in Sweden (Kiruna) and the Soviet Union. Titanium, phosphorus, and sulfur are often troublesome impurities in this kind of ore.

Magnetite series. A series of oxide minerals that crystallize in the isometric system and have the same atomic structure and similar physical properties, according to the chemical composition. It belongs to the spinel group and includes the following members:

Magnesioferrite $MgFe_2O_4$

Magnetite	$FeFe_2O_4$
Franklinite	$ZnFe_2O_4$
Jacobsite	$MnFe_2O_4$
Trevorite	$NiFe_2O_4$
Maghemite	Fe_2O_3

Malachite. A copper mineral and gem, $Ca_2CO_3(OH)_2$. Its color is green, and its luster varies. Occurring mostly in rounded forms, it is found in the Ural Mountains, South-West Africa, Rhodesia, the Congo, South Australia, and Arizona.

Maldonite. A mineral composed of gold and bismuth. It is soft and very heavy, and has a metallic-pink luster when fresh. It comes from Victoria (Australia).

Malladrite group. A group of halide minerals that crystallize in the hexagonal system and have the same atomic structure and a similar origin and occurrence. It includes the following members:

Malladrite	Na_2SiF_6
Bararite	$(NH_4)_2SiF_6$

Manganite. An orthorhombic mineral composed of manganese hydroxide, MnO(OH). Its crystals are black but may have a brown streak. This is one of the ores of manganese, mined in Germany, England, Michigan, and Nova Scotia.

Manganosite. An isometric mineral composed of manganese oxide, MnO. It appears glassy green, becoming black on exposure. Sweden and New Jersey are its localities.

Mansfieldite. An orthorhombic mineral of the scorodite-mansfieldite series in the variscite group. It occurs in Oregon as white, spongy and fibrous masses. The formula is $Al(AsO_4){\cdot}2H_2O$.

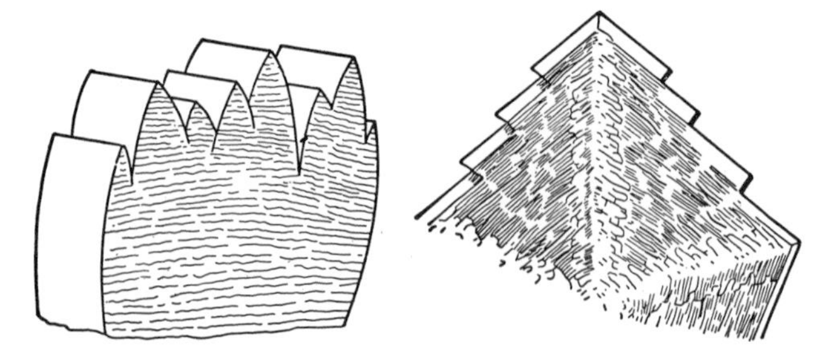

Cockscomb marcasite

Marcasite. An orthorhombic mineral composed of iron sulfide, FeS_2. It has a pale-brassy color, lighter than pyrite, which otherwise it closely resembles except in its crystal forms and fibrous structure. Marcasite has no commercial value. It occurs in England, central Europe, and the Tri-State district of Missouri, Kansas, and Oklahoma.

Margarite. A member of the brittle-mica group of minerals. It usually occurs in pink flakes. Important sources are Greece, Turkey, Massachusetts, Pennsylvania, and North Carolina. The crystallization is monoclinic, and the formula is $CaAl_2(Al_2Si_2O_{10})(OH)_2$.

Marialite. A tetragonal mineral of the scapolite series. It corresponds to the formula $(Na,Ca)_4Al_3(Al,Si)_3Si_6O_{24}(Cl,CO_3,SO_4)$.

Marsh ore. An old name for limonite, indicating a common occurrence.

Martite. Hematite occurring in octahedral crystals. These were formerly magnetite and have altered as pseudomorphs. Utah and Mexico are superior localities for specimens.

Mascagnite group. A group of sulfate minerals that crystallize in the orthorhombic system and have the same atomic structure. It includes the following members, which form a complete series with each other, as well as possibly taylorite which may be intermediate:

Mascagnite	$(NH_4)_2SO_4$
Arcanite	K_2SO_4

Massicot. A mineral composed of lead oxide, PbO. It is usually yellow, sometimes with a reddish tint. It is deposited around several volcanoes in Mexico, among other sources.

Matara diamond, Matura diamond. Old names for colorless zircon, indicating a Ceylonese locality for this diamondlike gem.

Matlockite group. A group of halide minerals that crystallize in the tetragonal system and have the same atomic structure and similar physical properties. It includes the following members, the last two of which form an incomplete series with each other:

Matlockite	PbFCl
Bismoclite	BiOCl
Daubreéite	BiO(OH,Cl)

Meadow ore. An old name for limonite and other minerals, indicating their occurrence.

Mealy zeolite. An old name for natrolite and mesolite, owing to the occasional mealy appearance of these zeolite minerals.

Meerschaum. The mineral sepiolite, used for pipes.

Meionite. A tetragonal mineral of the scapolite series. It has the formula $(Ca,Na)_4Al_3(Al,Si)_3Si_6O_{24}(Cl,CO_3,SO_4)$.

Melaconite. A triclinic mineral composed of copper oxide, CuO. It is the same as tenorite, a black mineral found in the upper levels of copper deposits.

Melanite. The black variety of andradite garnet. This mineral occurs in igneous rocks.

Melanterite. A monoclinic mineral composed of hydrous iron sulfate, $FeSO_4 \cdot 7H_2O$. It usually occurs in rounded forms which are green and glassy looking. Dozens of localities have been reported.

Melanterite group. A group of sulfate minerals that crystallize in the monoclinic system and have the same atomic structure and a similar origin and occurrence. They grade more or less into one another to make several natural series. The members are as follows:

Melanterite series	
Boothite	$CuSO_4 \cdot 7H_2O$
Pisanite	$(Fe,Cu)SO_4 \cdot 7H_2O$
Kirovite	$(Fe,Mg)SO_4 \cdot 7H_2O$

Melanterite series. A series of minerals in the melanterite group which grade completely into one another and have similar physical properties. It includes the following members:

Melanterite	$FeSO_4 \cdot 7H_2O$
Pisanite	$(Fe,Cu)SO_4 \cdot 7H_2O$
Kirovite	$(Fe,Mg)SO_4 \cdot 7H_2O$

Mendozite group. A group of sulfate minerals that crystallize in the monoclinic system and have the same atomic structure and a similar origin and occurrence. It includes the following members:

Mendozite	$NaAl(SO_4)_2 \cdot 11H_2O$
Kalinite	$KAl(SO_4)_2 \cdot 11H_2O$

Meneghinite. An orthorhombic mineral composed of lead-antimony sulfide, $Pb_{13}Sb_7S_{23}$. It has a fibrous look and a metallic-gray color. This is one of the feather ores. Italy is the most important locality.

Metatorbernite group. A group of phosphate-arsenate minerals that crystallize in the tetragonal system and have the same atomic structure and similar physical properties, especially radioactivity. The following members are included:

Metatorbernite	$Cu(UO_2)_2(PO_4)_2 \cdot 8H_2O$
Meta-autunite	$Ca(UO_2)_2(PO_4)_2 \cdot 2\text{–}6H_2O$
Metazeunerite	$Cu(UO_2)_2(AsO_4)_2 \cdot 8H_2O$

Metavariscite group. A group of phosphate minerals that crystallize in the monoclinic system and have the same atomic structure, probably grading into each other as a complete series. It includes the following members:

Metavariscite	$Al(PO_4) \cdot 2H_2O$
Metastrengite	$Fe(PO_4) \cdot 2H_2O$

Mexican. A name added to several gems to suggest a more valuable gem. Examples include *Mexican jade* and *Mexican onyx* for calcite.

Micaceous hematite, Micaceous iron ore. Names used for specular hematite, describing its habit.

Mica group. A group of silicate minerals that crystallize in the monoclinic system and have the same atomic structure and similar physical properties, especially elasticity. The following members are included, some of which grade into one another as partial series:

Muscovite	$KAl_2(AlSi_3O_{10})(OH)_2$
Paragonite	$NaAl_2(AlSi_3O_{10})(OH)_2$
Phlogopite	$KMg_3(AlSi_3O_{10})(OH)_2$
Biotite	$K(Mg,Fe)_3(AlSi_3O_{10})(OH)_2$
Lepidolite	$K_2Li_3Al_3(AlSi_3O_{10})(O,OH,F)_4$

Microcline. A triclinic potassium feldspar, $KAlSi_3O_8$. It also belongs to the alkali feldspars. The color is usually pink, white, or gray; green microcline is called amazonite or amazonstone. Large crystals and masses occur in pegmatites throughout the world.

Microlite. An isometric mineral composed of a complex tantalum oxide, $(Na,Ca)_2Ta_2O_6(O,OH,F)$. It is a member of the pyrochlore-microlite series. The color is usually yellow or brown. It has come from pegmatites and placers in New Mexico, Norway, Sweden, and Western Australia.

Milky quartz. Quartz made translucent or opaque by the presence of many liquid inclusions. *Gold quartz* is a variety of milky quartz.

Millerite. A hexagonal mineral composed of nickel sulfide, NiS. Its crystals are typically long and slender. Occurrences are scattered; they include Wales and various places in central Europe, the United States, and Canada.

Mimetite. A mineral of the pyromorphite series in the apatite group. The heavy, hexagonal crystals are usually rounded and barrel shaped. The luster is resinous to adamantine, and the color is usually yellow to brown. Leading sources include Germany, Great Britain, and South-West Africa. The formula is $Pb_5Cl(AsO_4)_3$.

Minium. A mineral composed of lead oxide, Pb_3O_4. It occurs in earthy, red masses. Germany is a common source.

Minnesotaite. A mineral similar to talc. It occurs in the iron deposits of Minnesota.

Mispickel. A miners' name of German origin, for arsenopyrite.

Mocha stone. An old name for moss agate.

Mock-lead. An English and American miners' name for sphalerite, from its frequent resemblance to galena, an ore of lead.

Molybdenite. A hexagonal mineral composed of molybdenum sulfide, MoS_2. It occurs in soft, metallic-gray sheets or masses, which can be cut with a knife. The chief ore of molybdenum, this mineral has its maximum concentration in Colorado (Climax), but it is scattered widely through the porphyry-copper deposits of the world.

Molybdenite group. A group of sulfide minerals that crystallize in the hexagonal system and have similar atomic structures and physical properties, except that tungstenite is heavier. The following members are included:

Molybdenite	MoS_2
Tungstenite	WS_2

Molybdic silver. An old name for wehrlite, based on wrong analyses.

Monazite. A monoclinic, rare-earth phosphate mineral. It occurs in brown, resinous-looking masses and as sand. It is the main source of thorium and comes from Brazil, India, Australia, South Africa, Florida, and North Carolina. The formula is $(Ce,La,Y,Th)PO_4$.

Montebrasite. A member of the amblygonite series of minerals. It is found principally as white masses having a vitreous or pearly luster. Pegmatite yields it in Sweden, South-West Africa, Brazil, Western Australia, Spain, South Dakota, and Maine. The crystallization is triclinic, and the formula is $(Li,Na)Al(PO_4)(OH,F)$.

Monticellite. A member of the olivine group of minerals. It is occasionally found in marble. The formula is $CaMg(SiO_4)$.

Montmorillonite. A mineral of the montmorillonite group of clay minerals. It swells in water and is the chief constituent of the rock called bentonite. The formula is $Na(Al,Mg)_2Si_4O_{10}(OH)_2$.

Montmorillonite group. A group of clay minerals characterized by swelling in water. It includes the following members:

Montmorillonite	$Na(Al,Mg)_2Si_4O_{10}(OH)_2$
Nontronite	$NaFe_2(Al,Si)_4O_{10}(OH)_2$
Saponite	$CaMg_3(Al,Si)_4O_{10}(OH)_2$
Hectorite	$Na(Mg,Li)_3Si_4O_{10}(F,OH)_2$
Sauconite	$Ca,Na(Zn,Mg,Al,Fe)_3(Al,Si)_4O_{10}(OH)_2$
Stevensite	$CaMg_3Si_4O_{10}(OH)_2$

Moonstone. A gem variety of feldspar, especially oligoclase (plagioclase) but also adularia (orthoclase) or albite (plagioclase). It shows a delicate, bluish radiance. *Pink moonstone* is the mineral scapolite.

Morganite. The pink to rose, gem variety of the mineral beryl, colored by lithium. It typically occurs with pink tourmaline and spodumene, both also colored by lithium. Madagascar is the best locality; California has yielded good material.

Morion. A dark-brown variety of quartz. It is similar to smoky quartz.

Moschellandsbergite. A metallic-white mineral composed of a compound of silver and mercury, Ag_2Hg_3. It occurs in isometric crystals and masses. The best crystals have come from Germany, but Sweden and France have yielded good specimens.

Moss agate. Chalcedony marked by dendritic patterns. These are made by manganese oxide in the form of the mineral pyrolusite, or by iron oxide, or by the mineral chlorite. The color is black, brown, green, or blue. Among the numerous descriptive names used for this gem are *Mocha stone, flower agate, plume agate, scenic agate, landscape agate, seaweed agate,* and *tree agate.* The western United States is the principal source of moss agate.

Mountain blue. An old name for azurite and chrysocolla, describing their color.

Mountain cork. A light, corklike asbestos which floats on water.

Mountain green. An old name for malachite and chrysocolla, because of their color.

Mountain leather. Asbestos consisting of thin, flexible sheets, composed of interlaced fibers.

Mountain mahogany. An old name for brown obsidian, referring to its appearance.

Mountain tallow. An old name for hatchettite, describing its fatty appearance.

Mountain wood. A compact asbestos resembling wood.

Muller's glass. An old name for hyalite opal.

Muscovite. The white mica. Green and various other colors are seen in thick specimens of this monoclinic mineral. Muscovite is a hydrous silicate of aluminum and potassium, $KAl_2(AlSi_3O_{10})(OH)_2$. Huge crystals grew in pegmatite in India, Ontario, and localities in the Rocky and Appalachian Mountains. Metamorphic rocks are another important source.

Nacrite. A clay mineral of the kaolin group. Its formula is $Al_2Si_2O_5(OH)_4$.

Nagyagite. A monoclinic, telluride mineral, $Pb_5Au(Te,Sb)_4S_{5-8}$. It is soft and heavy, gray, and metallic looking. This uncommon mineral has been reported from various localities in Rumania, Western Australia, New Zealand, Fiji, Japan, Canada, and the United States.

Nantokite group. A group of halide minerals that crystallize in the isometric system and have the same atomic structure and a similar origin and occurrence. It includes the following members:

Nantokite	CuCl
Miersite	AgI
Marshite	CuI

Native antimony. A metallic-white mineral, usually occurring in irregular masses. Fine specimens have come from Borneo and Australia. Most antimony of commerce comes from stibnite and other minerals.

Native arsenic. A mineral that occurs, when fresh, in metallic-white, rounded masses; these soon tarnish dark gray. Most of the localities are in central Europe, but native arsenic is rather uncommon. The arsenic of commerce is obtained as a byproduct of ore smelting.

Native bismuth. A mineral occurring in metallic-white networks or branching shapes. Australia and Bolivia have the main deposits. The bismuth of commerce is obtained as a byproduct of other mining or of ore smelting.

Native copper. A heavy, soft, metallic-red mineral that occurs in crude crystals, as wire, and in irregular masses. It is bright-rose colored when fresh, tarnishing readily. The unique deposits of Michigan (Keweenaw Peninsula) have yielded pieces as large as 420 tons. Besides the native mineral, copper is obtained from minerals that are compounds of copper (chalcocite, bornite, chalcopyrite, tetrahedrite, enargite, antlerite) and from other minerals, such as pyrite, in which copper is a chemical impurity.

Native gold. One of the most beautiful of minerals when well crystallized. Soft, metallic-yellow nuggets have been found up to the size of the Welcome Stranger, from Ballarat, Victoria (Australia), which weighed 2,280 ounces. Gold is always alloyed with a varying proportion of silver. Its density depends on its purity, reaching a maximum when free of silver. Besides the native mineral, the gold of commerce is obtained from some telluride minerals and base-metal minerals, such as pyrite and chalcopyrite, in which gold is a chemical impurity. Gold is world-wide in distribution, but 18,000 tons, valued at 14 billion dollars, have come from the Witwatersrand district, South Africa. The total world store now amounts to about 50,000 tons.

Native lead. A very rare mineral, occurring in heavy, soft, metallic-gray crystals which tarnish quickly. Most of it comes from Sweden. Rounded masses, thin plates, and wires have come from the Soviet Union, Madeira, Mexico, New Jersey, and Idaho.

Native mercury. An extraordinarily heavy liquid, becoming solid at –39°C, when it turns to metallic-white crystals. The rare mineral occurs as separate, metallic-gray drops associated with cinnabar in volcanic regions or near hot springs in Texas, California, Spain, Italy, Yugoslavia, and Germany.

Native palladium. A heavy, metallic-gray mineral. It is rare, though found in Columbia, Brazil, the Soviet Union, and South Africa.

Native platinum. This very heavy mineral occurs mostly in metallic-gray grains, scales, and nuggets. Besides the native mineral, the platinum of commerce is obtained from sperrylite. Production of this metal is chiefly in Canada, South Africa, and the Soviet Union. The platinum group of metals also includes iridium (Ir), palladium (Pd), osmium (Os), rhodium (Rh), and ruthenium (Ru), all of which occur in intimate association. They are used in jewelry and for other industrial purposes.

Native sedative salt. An old name for sassolite (boric acid), from its early use in medicine.

Native selenium. A soft, metallic-gray mineral, occurring as flexible crystals and droplets. It has been produced by mine fires in Arizona and Czechoslovakia. The selenium of commerce is obtained mostly as a byproduct of smelter operations.

Native silver. A heavy, and soft, metallic-white mineral, which occurs in crude crystals, as wire, and in nuggets and masses. It tarnishes easily to gray or black. Superb specimens used to come from Kongsberg, Norway. Huge pieces are known from Ontario (Cobalt) and Colorado (Aspen). Besides the native mineral, and its occurrence in native gold, the silver of commerce comes from minerals that are compounds of silver and also from base-metal minerals, such as galena, in which silver is a chemical impurity. Latin America is the greatest source of silver, chiefly Mexico, Peru, and Bolivia.

Native sulfur. A soft, light mineral, occurring in enormous quantities in the cap rock of salt domes along the Gulf of Mexico. Sulfur is found with most volcanoes, and in beds of rock containing gypsum, limestone, clay, or bituminous material. Excellent yellow crystals come from Sicily. Besides the native mineral, the sulfur of commerce is obtained from pyrite and from natural gas.

Native tellurium. A metallic-white mineral, occurring as masses and crystals. It is found especially in Japan and Colorado, but also in Rumania, Western Australia, Asia Minor, and Nevada. The tellurium of commerce is obtained as a byproduct of smelter operations.

Native tin. A soft and heavy mineral, found in rounded, metallic-white grains in New South Wales (Australia). The tin of commerce is obtained from cassiterite and stannite.

Native zinc. A soft and heavy, metallic-white mineral reported from various places. However, it is either extremely rare or actually nonexistent. The zinc of commerce is obtained from sphalerite and other minerals.

Natrolite. A member of the zeolite family of minerals. Its delicate, radiating, needlelike crystals are characteristic. Sometimes this vitreous, white mineral is tinged yellow to red, and it may be compact enough to carve, as is the material from Pennsylvania. The crystallization is monoclinic, and the formula is $Na_2(Al_2Si_3O_{10}) \cdot 2H_2O$.

Natromontebrasite. A member of the amblygonite series of minerals. The white masses have a vitreous or pearly luster. The source is in pegmatite in Colorado. The crystallization is triclinic, and the formula is $(Na,Li)Al(PO_4)(OH,F)$.

Naumannite. A mineral composed of silver selenide, Ag_2Se. It occurs in metallic-black crystals and granular masses. Substantial amounts have come from Idaho, and some from Germany and Argentina.

Needle-ore. A miners' name for aikinite, indicating its frequent acicular habit.

Needle-tin ore. An old miners' name for long crystals of cassiterite, the ore of tin.

Nepheline. A member of the feldspathoid group of minerals. It occurs as vitreous, colorless, hexagonal crystals and greasy, gray masses. The Soviet Union, Norway, and South Africa contain large bodies, and other localities are in Ontario, Maine, New Jersey, Arkansas, and Mount Vesuvius. The massive variety is called eleolite. The formula is $(Na,K)AlSiO_4$.

Nephrite. One of the two forms of true jade. It is a member of the amphibole family of minerals, in the tremolite-actinolite series—being the compact and tough variety of either one, according to whether it is white (tremolite) or green (actinolite). Nephrite is the less valuable of the two jade materials because of its inferior luster and range of color. It is of metamorphic origin and comes from China, New Zealand, Wyoming, and Alaska.

Niccolite. A hexagonal mineral composed of nickel arsenide, NiAs. It has a copper-red color. Rounded masses occur in Japan, Germany, and France. Ontario (Cobalt) is a locality for vein niccolite.

Niccolite group. A group of sulfide-arsenide-antimonide minerals that crystallize in the hexagonal system, except pentlandite, which is isometric. The first three have the same atomic structure. The following members are included:

Pyrrhotite	$Fe_{1-x}S$
Niccolite	NiAs
Breithauptite	NiSb
Millerite	NiS
Pentlandite	(Fe,Ni)S

Nickel bloom. An old name for annabergite, because of its metal content and surface coatings.

Nickel glance. An old name for gersdorffite, indicating its chemical composition and bright, metallic luster.

Nickel-iron. A natural alloy of nickel and iron occurring in meteorites. It appears in iron meteorites as kamacite, as taenite, and as an intergrowth of the two, called plessite. Commercial nickel is obtained from pentlandite and garnierite.

Nickel vitriol. An old name for morenosite, owing to its metal content and sulfate composition.

Niter. An orthorhombic, nitrate mineral, KNO_3. It is found as vitreous, white crusts or silky, needlelike crystals. These have a salty, cooling taste. It has come from various soils in a number of places and, being highly valued for its use in gunpowder, was also collected from the walls and ceilings of cellars and shallow caves.

Nontronite. An iron-bearing member of the montmorillonite group of clay minerals. When pure, it is greenish yellow. The formula is $NaFe_2(Al,Si)_4O_{10}(OH)_2$.

Noselite. A blue, isometric mineral of the feldspathoid group. The formula is $Na_4Al_3Si_3O_{12}\cdot SO_4$.

Occidental. A name added to several gems to indicate a gem less valuable than its Oriental counterpart, which is often considered the most valuable. Examples include *Occidental cat's-eye* for cat's-eye quartz, *Occidental topaz* for citrine quartz, and *Occidental amethyst* for real amethyst.

Occidental turquoise. An old name for odontolite, indicating its resemblance to turquoise.

Ocher. An old name for various minerals having an earthy luster, such as *lead ocher* (massicot), *red ocher* (hematite), *tungstic ocher* (tungstite), and *black cobalt ocher* (wad).

Octahedral copper ore. An old name for cuprite, from its crystal form and use.

Octahedral iron ore. An old name for magnetite, indicating its crystal form and use.

Octahedrite. The less preferred name for anatase.

Odontolite. Fossil bone and teeth colored blue by the mineral vivianite. It looks like turquoise.

Oligoclase. A member of the plagioclase feldspar series of minerals. It consists of 70 to 90 percent albite and the rest anorthite. Most moonstone is oligoclase, and sunstone is another gem variety.

Olivine. A rock-forming and gem mineral. It is a member of the olivine series in the olivine group. The name is used either for the entire series or for the intermediate, common variety which is typically green, vitreous, and granular. It occurs as grains in dark igneous rocks and as separate bodies of its own (called dunite), sometimes also in marble, and even in meteorites. Crystals are found at Mount Vesuvius; granular masses, in volcanic bombs in Germany and Arizona; and large bodies of rock, in New Zealand and North Carolina. Olivine is used to make refractory bricks. The transparent, gem variety is known as peridot. Chrysolite is another name for olivine.

Olivine group. A group of silicate minerals that crystallize in the orthorhombic system and have the same structure and similar physical properties. It includes the following members:

Olivine series	
Monticellite	$CaMg(SiO_4)$
Tephroite	$Mn_2(SiO_4)$
Larsenite	$PbZn(SiO_4)$

Olivine series. A series of orthorhombic, silicate minerals in the olivine group, which grade completely into one another. It includes the following members:

Olivine	$(Mg,Fe)_2(SiO_4)$
Forsterite	$Mn_2(SiO_4)$
Fayalite	$Fe_2(SiO_4)$

Olivenite group. A group of arsenate-phosphate minerals that crystallize in the orthorhombic system and have the same atomic structure and a similar origin and mode of occurrence. They grade somewhat into one another to form several partial series. The following members are included:

Olivenite	$Cu_2(AsO_4)(OH)$
Libethenite	$Cu_2(PO_4)(OH)$
Adamite	$Zn_2(AsO_4)(OH)$

Onegite. A name given to amethyst having needlelike inclusions.

Onyx. Agate showing straight and wide bands of contrasting colors. It is especially suitable for carving into cameos. Sardonyx is a special variety. Most of the onyx of commerce is dyed.

Onyx marble. Compact, banded calcite or aragonite. Also known as *Mexican onyx,* it is used for decorative objects and comes mostly from Mexico.

Opal. A leading gem mineral, often of great beauty. It is composed of hydrous silica, $SiO_2 \cdot nH_2O$, and is one of the few amorphous minerals. The luster is vitreous or resinous. Common opal lacks the colorful internal reflections caused by the diffraction of light, and its over-all color is usually milky, although some other color may predominate. Precious opal includes varieties known as *white opal, black opal,* and *fire opal. Opalized wood* may contain either *common* or *precious opal. Hyalite, lechosos opal, harlequin opal,* and *geyserite* are special varieties of opal. The choicest gem opal comes from Australia, Nevada, Czechoslovakia, Mexico, and Honduras.

Opalite. A name given to impure common opal.

Opalized wood. Pertified wood that has been filled or replaced by opal. Some of it is gemmy, but most is not.

Orangite. The orange-colored variety of the mineral thorite.

Oriental. A name given to a number of gems to indicate the most valuable kind of a given appearance. Examples include *Oriental amethyst* for purple sapphire (corundum), *Oriental cat's-eye* for cat's-eye chrysoberyl, *Oriental emerald* for green sapphire (corundum), and *Oriental topaz* for yellow sapphire (corundum).

Orpiment. A monoclinic mineral composed of arsenic sulfide, As_2S_3. It is yellow and shows good cleavage. Orpiment is often accompanied by red realgar, making a colorful association.

Orthite. The name used in Europe for the mineral allanite.

Orthoclase. A monoclinic potassium feldspar, $KAlSi_3O_8$. It also belongs to the alkali feldspars. The color is usually pink, white, or gray. Large crystals and masses occur in pegmatite throughout the world. A clear, colorless variety called adularia includes some gem moonstone.

Ottrelite. A member of the brittle-mica group of minerals. It is found in metamorphic rocks. The crystallization is monoclinic, and the formula is $(Fe,Mn)(Al,Fe)_2Si_3O_{10} \cdot H_2O$.

Oxidulated copper. An old name for cuprite, referring to its chemical composition.

Oxidulated iron. An old name for magnetite, alluding to its chemical composition.

Pagoda stone. A popular name for agalmatolite, from its use in carved Oriental ornaments. It is also used in place of fossil limestone and figured agate for decorative purposes.

Paragonite. A sodium-bearing member of the mica group of minerals. It looks like muscovite.

Parawollastonite. A calcium silicate mineral, $CaSiO_3$, found in blocks of limestone thrown from volcanoes. Mount Vesuvius is the best locality.

Peacock ore. An English miners' name for bornite and chalcopyrite, especially bornite, owing to the purple-blue and bronzy-iridescent color of its tarnish.

Pea ore. A popular name for small, rounded masses of limonite (an iron ore), because of their shape and use.

Pearl. A gem secreted by a mollusk. It is composed of calcium carbonate (as calcite and aragonite) and conchiolin (an organic substance). Marine and river waters throughout the world have pearl fisheries.

Pearl spar. An old name for dolomite and other similar minerals, referring to their bright luster on the cleavage surface.

Pectolite. A mineral of the pyroxenoid group. It is characterized by needlelike crystals that are dangerous to handle. They are white and vitreous or even silky looking but sharp and brittle. Pectolite is usually found in cavities in basalt with other late minerals, such as zeolites, calcite, and prehnite. New Jersey is a chief locality. The crystallization is triclinic, and the formula is $Ca_2NaHSi_3O_9$.

Penninite. A member of the chlorite group of minerals.

Pentlandite. An isometric mineral composed of iron-nickel sulfide, $(Fe,Ni)_9S_8$. A bronze-colored mineral, it is almost always intermixed with pyrrhotite, as in Ontario (Sudbury), where it is the world's main source of nickel.

Periclase. An isometric mineral composed of magnesium oxide, MgO. It is a white, glassy-looking mineral found at Mount Vesuvius and other localities where high temperatures have existed.

Periclase group. A group of oxide minerals that crystallize in the isometric system and have the same atomic structure. The following members are included:

Periclase	MgO
Bunsenite	NiO
Manganosite	MnO

Peridot. The gem variety of the mineral olivine. Its bottle-green color is sometimes very attractive and it has been popular since ancient times. An island in the Red Sea is perhaps the classic locality, and other specimens come from Burma and from surface gravel in Arizona and New Mexico.

Peristerite. A gem variety of albite (plagioclase) feldspar, showing a play of colors.

Perovskite. An isometric mineral composed of calcium-titanium oxide, $CaTiO_3$. It is usually black and nearly metallic looking. Colorado contains a large deposit.

Perthite. A mineral intergrowth of plagioclase feldspar and potash feldspar.

Petalite. A member of the feldspathoid group of minerals. It occurs usually in platy masses which are vitreous and colorless, white, or gray. Large deposits are mined for lithium in Southern Rhodesia and South-West Africa. Its crystallization is monoclinic, and the formula is $Li(AlSi_4O_{10})$.

Petoskey stone. Named after Petoskey, Michigan, the locality for a particularly attractive variety of petrified coral.

Petrified wood. Wood that has been replaced by silica or other mineral matter. Often merely the pore spaces have been partly or completely filled, but both filling and replacement usually occur. The more gemmy varieties include agatized wood, jasperized wood, and opalized wood. The United States, Canada, and Argentina are among the principal sources of petrified wood.

Petzite. A heavy mineral composed of silver-gold telluride, $(Ag,Au)_2Te$. Metallic-gray masses, probably isometric in crystallization, occur in Rumania, Western Australia, Colorado, California, and Ontario.

Phenacite. A gem mineral of the phenacite group, composed of beryllium silicate, Be_2SiO_4. It is vitreous and usually colorless, resembling clear quartz. Its rhombohedral crystals and hardness ($7\frac{1}{2}$ to 8) help to identify it. A rare mineral, phenacite occurs in the Ural Mountains, Brazil (Minas Geraes), and Colorado (Mount Antero).

Phenacite group. A group of silicate minerals that crystallize in the hexagonal system and have the same structure. It includes the following members:

Phenacite	$Be_2(SiO_4)$
Willemite	$Zn_2(SiO_4)$

Phillipsite. A member of the stilbite group of minerals, in the zeolite family. The crystallization is monoclinic, and the formula is $(Al_3Si_5O_{16})\cdot 6H_2O$.

Phlogopite. The brown mica. In addition to the fundamental mica formula, it contains magnesium, becoming $KMg_3(AlSi_3O_{10})$

$(OH)_2$. It usually occurs in metamorphic rocks. Ontario, Quebec, New York, Madagascar, Ceylon, Switzerland, Sweden, and Finland have important deposits.

Phosgenite. A tetragonal, carbonate mineral, $Pb_2Cl_2CO_3$. It has a bright, nonmetallic luster and is generally brown or white. Especially good specimens have come from England, Poland, and Colorado.

Picotite. Black spinel.

Picromerite group. A group of sulfate minerals that crystallize in the monoclinic system and have the same atomic structure and probably a similar occurrence. It includes the following members:

Picromerite	$K_2Mg(SO_4)_2 \cdot 6H_2O$
Cyanochroite	$K_2Cu(SO_4)_2 \cdot 6H_2O$
Boussingaultite	$(NH_4)_2Mg(SO_4)_2 \cdot 6H_2O$

Piedmontite. A manganese-bearing member of the epidote-clinozoisite series of minerals. Its color is violet red. The formula is $Ca_2(Al,Fe,Mn)Al_2O(SiO_4)(Si_2O_7)(OH)$.

Pigeonite. A member of the pyroxene family of minerals. It is a complex silicate intermediate between augite and clinoenstatite. It is found in dark volcanic rocks.

Pinite. Mica altered from other minerals. It is usually similar to muscovite.

Pink moonstone. A gem variety of the mineral scapolite.

Pipestone. A popular name for catlinite, referring to its use for Indian pipes.

Pistacite. Another name for the mineral epidote, referring to its pistachio-green color.

Pitchblende. The impure variety of uraninite. It is the main source of uranium and, therefore, of nuclear energy. It is usually pitchy in appearance, but is sometimes either dull or glossy and occasionally banded or fibrous.

Pitch iron ore, Pitch ore. Old names for pitticite or general hydrous iron sulfate-arsenate minerals of uncertain identity.

Plagioclase series. A series of triclinic, silicate minerals in the feldspar group, which grade completely into one another. They are arbitrarily divided into six members, according to the relative proportions of the so-called "end members" albite, $NaAlSi_3O_8$, and anorthite, $CaAl_2Si_2O_8$. Because of this composition, they are also known as the soda-lime feldspars. They are generally white or gray, or less often reddish, yellowish, or greenish. A characteristic feature is the presence of striations on certain cleavage surfaces.

They are named as follows, the first three being *sodic plagioclase,* the last three being *calcic plagioclase:*

Albite	90-100	percent	albite
Oligoclase	70- 90	"	"
Andesine	50- 70	"	"
Labradorite	30- 50	"	"
Bytownite	10- 30	"	"
Anorthite	0- 10	"	"

Plagionite. A monoclinic mineral composed of lead-antimony sulfide, $Pb_5Sb_8S_{17}$. Its crystal and masses are metallic black. Germany, France, and Bolivia are reported localities.

Plagionite group. A group of sulfosalt minerals that crystallize in the monoclinic system and are similar in atomic structure and physical properties. Although they have compositions similar to the so-called feather ores (jamesonite, etc.), they occur in solid crystals that are short prisms or tabular instead of fibrous or needlelike. The following members are included:

Fuloppite	$Pb_7Sb_8S_{15}$
Plagionite	$Pb_5Sb_8S_{17}$
Heteromorphite	$Pb_7Sb_8S_{19}$
Semseyite	$Pb_9Sb_8S_{21}$

Plasma. Green chalcedony quartz. It often has white or yellow spots on it.

Platiniridium. A very rare, precious-metal mineral. It is metallic white, hard, and extremely heavy. The composition is Pt,Ir. Brazil, the Soviet Union, and Burma are known localities.

Platinum group. A group of native element minerals that crystallize in the isometric and hexagonal systems and have similar physical properties. They occur together and are often difficult to separate. The following members are included:

Platinum	Pt
Palladium	Pd
Platiniridium	Ir,Pt
Aurosmiridium	Ir,Os,Au
Iridosmine series	
Allopalladium	Pd

Plumbago. An old name for galena and especially graphite, coming from the Latin word for lead, which these minerals either contain (galena) or resemble (graphite).

Plumbic ocher. An old name for massicot, from its lead content and earthy look.

Plumbogummite group. A group of phosphate minerals that probably crystallize in the hexagonal system and have the same atomic structure. A number of them have a similar occurrence. The following members are included:

Plumbogummite	$PbAl_3(PO_4)_2(OH)_5H_2O$
Gorceixite	$BaAl_3(PO_4)_2(OH)_5H_2O$
Goyazite	$SrAl_3(PO_4)_2(OH)_5H_2O$
Crandallite	$CaAl_3(PO_4)_2(OH)_5H_2O$
Deltaite	$Ca(Al_2Ca)(PO_4)_2(OH)_5H_2O$
Florencite	$CeAl_3(PO_4)_2(OH)_6$
Dussertite	$BaFe_3(AsO_4)_2(OH)_5H_2O$

Plume agate. Moss agate showing a plumelike pattern.

Plumose antimonial ore. An old name for jamesonite and stibnite, because of their sometimes feathery habit and antimony content.

Plush copper ore. A popular name for the chalcotrichite variety of cuprite, from its rich appearance and use.

Polianite. The variety of pyrolusite that occurs in hard crystals.

Pollucite. An isometric, silicate mineral, $Cs_4Al_4Si_9O_{26}\cdot H_2O$. It occurs usually in pegmatite.

Polybasite. A monoclinic mineral composed of silver-antimony sulfide, $Ag_{16}Sb_2S_{11}$. It has a steel-gray color and shows triangular markings on the top crystal faces. A number of silver mines in the United States, Mexico, Chile, and Germany produce polybasite as an ore.

Polybasite group. A group of sulfosalt minerals that crystallize in the monoclinic system and have the same atomic structure and similar physical properties. It includes the following members:

Polybasite	$(Ag,Cu)_{16}Sb_2S_{11}$
Pearceite	$(Ag,Cu)_{16}As_2S_{11}$

Polyhalite. A potash mineral formed by evaporation. It is gray or red and has resinous luster and a bitter taste. The main sources are New Mexico, Texas, Germany, and Austria. The crystallization is triclinic, and the formula is $K_2Ca_2Mg(SO_4)_4\cdot 2H_2O$.

Poppy stone. A popular name for orbicular jasper, because of its flowerlike patterns.

Potash mica. Muscovite, so called from its chemical composition.

Potters' ore. A name for galena occurring in coarse-grained lumps large enough to be used to glaze pottery.

Prase. Green jasper. Good specimens have come from Germany.

Prehnite. An attractive, light-green mineral occurring mostly in rounded masses having a series of flat, vitreous surfaces. It occurs chiefly in lava rocks and in association with calcite, zeolites, datolite, and pectolite. New Jersey and the Lake Superior copper district contain good prehnite localities.

Prismatic arsenical pyrites. An old name for loellingite and probably other minerals, because of the long crystals and the chemical composition.

Prismatic arsenious acid. An old name for claudetite, because of the crystal shape and chemical composition.

Prismatic iron pyrites. An old name for marcasite, from the occasional length of its crystals and its resemblance to pyrite.

Prismatic manganese-ore. An old name for pyrolusite, referring to the shape of its crystals and its use.

Prochlorite. A member of the chlorite group of minerals.

Proustite. Light ruby silver. It is a hexagonal mineral composed of silver-arsenic sulfide, Ag_3AsS_3, having a red color that is lighter than that of pyrargyrite. Associated with, but less common than, pyrargyrite, it is found in Idaho, central Europe, and Latin America.

Pseudo-galena. An old name for sphalerite, from its occasional resemblance to galena.

Pseudomanganite. An older name for pyrolusite, indicating its resemblance to manganite.

Pseudowollastonite. A high-temperature, calcium silicate mineral, $CaSiO_3$.

Psilomelane. An orthorhombic mineral composed of barium-manganese oxide, $(Ba,H_2O)_2Mn_5O_{10}$. It is a black, massive mineral with rounded shapes. As an ore of manganese, it occurs in Germany, France, and India.

Pumpellyite. A monoclinic mineral related to epidote. It is brown or green, occurring in fibers or tablets in metamorphic rock. The formula is $Ca_4(Mg,Fe,Mn)(Al,Fe,Ti)_5(OH)_3(Si_2O_7)_2(SiO_4)_2 \cdot 2H_2O$.

Purple copper ore. An old name for bornite, in reference to the color of its tarnish and its use.

Pyramidal manganese ore. An old name for hausmannite, because of the shape of its crystals and its use.

Pyramidal zeolite. An old name for apophyllite, formerly grouped with the zeolite family, describing the crystal form of this mineral.

Malachite (green), azurite (blue), and tenorite (black).

This trio of minerals from Idaho are natural companions, as they are all oxidation products of copper.

Cerussite crystals.

From Tiger, Arizona, comes this choice group. The grillwork of the interior suggests a steel-girder building under construction. This open structure represents one of cerussite's characteristic modes of crystal growth.

Smithsonite.

The botryoidal form is the usual habit of this zinc carbonate. This is one of its most common hues. From Kelly, New Mexico.

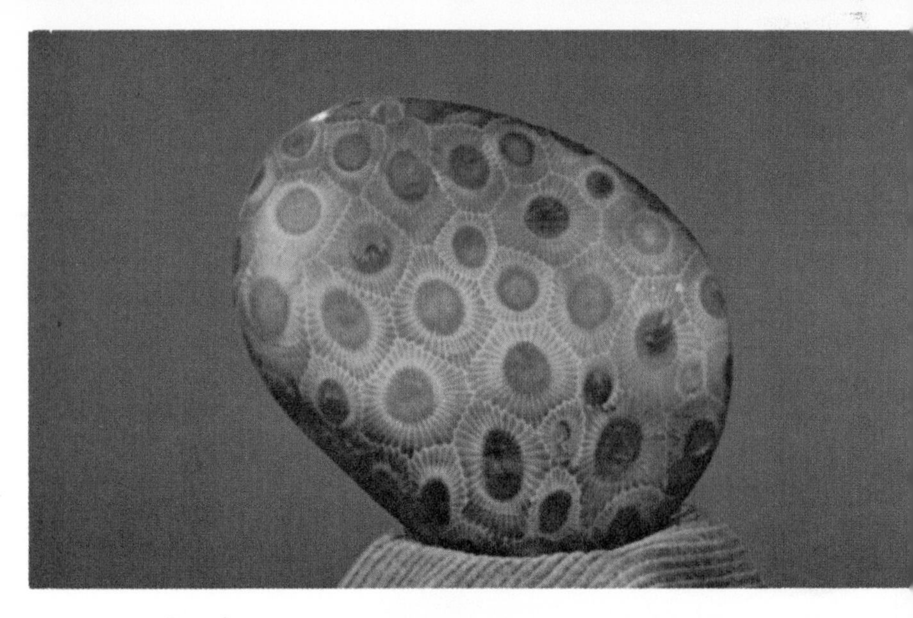

Petoskey stone. A petrified coral from Traverse City, Michigan. The rounding of this pebble by water abrasion shows the hexagonal pattern of this coral to greater advantage.

Golden, iridescent calcite from Mexico. The gleam of these masses of sharply pointed crystals is enhanced by an iridescence.

Calcite in rhyolite geode.

Protected in its own roomy cell—the hollow, crystal-lined cavity of the geode—this calcite crystal grew to a perfect termination. Sparkling from the depths are quartz crystals. From Blythe, California.

Nailhead calcite.

Blunt crystals, squatting atop other crystals, give this calcite its name and represent one of the many odd forms of this mineral. Ohio is the source of this specimen. Calcite occurs in a wider variety of forms than any other mineral (as you have probably noted by now).

Colemanite.

One of the beautiful borate minerals from Boron, California. The sharp, brilliant crystals make a group handsome enough for any cabinet.

Colemanite with calcite.

These two have grown together at Boron, California. The colemanite is clear.

PHOSPHATES

Apatite crystal. A well-formed, hexagonal crystal with a pyramidal termination; from Huddersfield, Quebec. Note the typically crackled appearance. The original name was *asparagus stone,* after the yellowish-green variety from Murcia, Spain.

Ludlamite.

These quiet-green, tabular crystals from the Blackbird district, Lemhi County, Idaho, appear to be sprinkled with tiny metallic crystals.

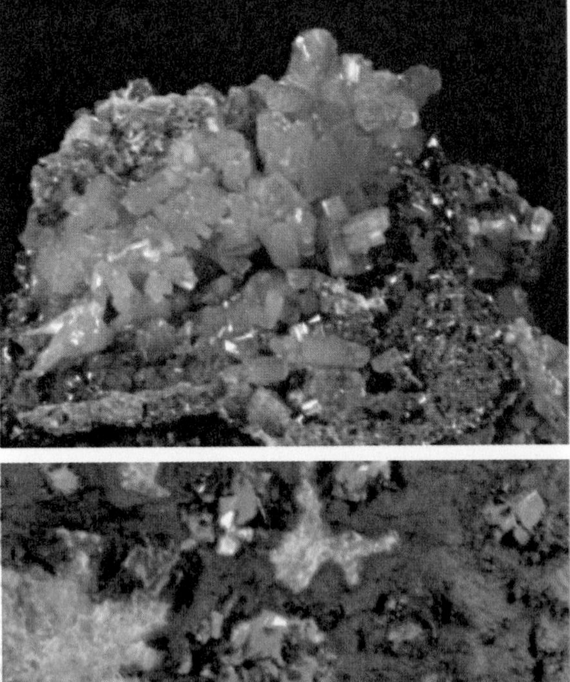

Pyromorphite.

White clusters bursting from a cavity in galena show that pyromorphite is a secondary mineral, often found in lead deposits. These crystals come from Germany.

Torbernite on conglomerate.

These bright, emerald-green crystals derive their color from copper and their radioactivity from uranium. From Mexico.

Apatite crystal. The matchless form, color, clarity, and luster of this crystal from Mexico make it excellent gem material.

Autunite crystals. From a recently discovered locality near Spokane, Washington, come these bundles of radioactive mineral. The quality is outstanding.

Brazilianite. An example of one of the most recently discovered gem minerals. Brazilianite, of course, comes from Brazil.

VANADATES

Descloizite. An unfolding of substantial crystals characterizes this fine specimen from Africa.

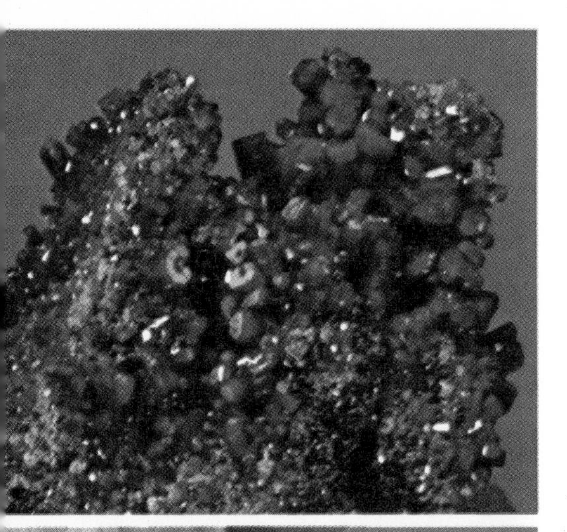

Vanadinite.

In these brilliant, brownish-red crystals from Mexico we see a collectors' favorite. The hollow, cavernous structure of the six-sided crystals, a fairly familiar feature, can be seen at the center of the picture.

Vanadinite on calcite.

Standing out abruptly from their surroundings of calcite rhombs, these golden crystals of vanadinite are sharply defined and terminated hexagonal prisms. This unusual group comes from Mexico, the country where vanadinite was first observed.

Hewettite crystals with calcite.

The enmeshed bundles look almost like vegetation. This interesting specimen is from Utah.

ARSENATES

Conichalcite with calcite. This combination from Mexico is perhaps more interesting than attractive; note the intricate crystal growth.

Adamite.

The brownish-orange matrix is limonite. From Mexico.

Adamite on limonite.

This colorful specimen is also from Mexico.

Mimetite. One of the fine specimens from Africa; note the glistening, granular coating on its long crystals.

Mimetite. These crystals closely resemble pyromorphite in crystal habit and chemical composition. From England.

Erythrite. One can almost taste this extraordinary cluster of purplish crystals. From Mexico.

Linarite and caledonite. This specimen well portrays one of the ways in which crystals of two chemically different minerals grow side by side. Linarite is the darker mineral. From Tiger, Arizona.

Selenite twin.

This odd-looking specimen from Oklahoma is called a *swallow-tail twin.* Note the smaller twin offshoots.

Celestite crystals.

Here is the heavenly blue that gives celestite its name. Ohio is the source of this cluster.

Brochantite and cerussite.

The velvety, green brochantite has formed upon the branching crystals of cerussite. This beautiful combination is from Tiger, Arizona.

Pyrargyrite. Dark ruby silver. It is a hexagonal mineral composed of silver-antimony sulfide, Ag_3SbS_3, having a deeper red color than proustite, with which it associates. Idaho has yielded important amounts of pyrargyrite, which also comes from central Europe and Latin America.

Pyrite. The most common sulfide and one of the most common metallic minerals. It is an isometric mineral composed of iron sulfide, FeS_2. Having a brassy color, it is one of the fool's gold minerals. Its brittleness and hardness (6 to 6½) will distinguish it from the other fool's golds and true gold. Crystals are often very handsome, as are those from Colorado (Leadville). Masses of large size occur in Virginia, Quebec, and Spain. Pyrite is an important ore of gold and copper, which are present as chemical impurities, and it is sometimes a source of iron, sulfur, and sulfur dioxide.

Pyrite group. A group of sulfide-arsenide-selenide minerals that crystallize in the isometric system and have the same atomic structure and rather similar physical properties, although hauerite does not have a metallic luster. The following members are included:

Pyrite	FeS_2
Bravoite	$(Ni,Fe)S_2$
Laurite	RuS_2
Sperrylite	$PtAs_2$
Hauerite	MnS_2
Penroseite	$(Ni,Cu,Pb)Se_2$

Pyritous copper. An old name for chalcopyrite, because of its resemblance to pyrite and its chemical composition.

Pyrochlore. An isometric mineral composed of a complex columbium oxide, $NaCaCb_2O_6F$. It is a member of the pyrochlore-microclite series. The color is brown to black. Norway, Sweden, and Ontario have yielded pyrochlore from pegmatite and other rocks.

Pyrochlore-microlite series. A series of oxide minerals that crystallize in the isometric system and have the same atomic structure and similar physical properties, according to the chemical composition. It includes the following members:

Pyrochlore	$NaCaCb_2O_6F$
Microlite	$(Na,Ca)_2Ta_2O_6(O,OH,F)$

Pyrolusite. A tetragonal mineral composed of manganese oxide,

MnO_2. It is soft enough to mark the fingers black. Crystals, called polianite, are hard (6 to 6½). Pyrolusite is the substance that makes most of the dendrites seen on rock surfaces. It is an important ore of manganese and is mined in the Soviet Union, India, South Africa, and elsewhere.

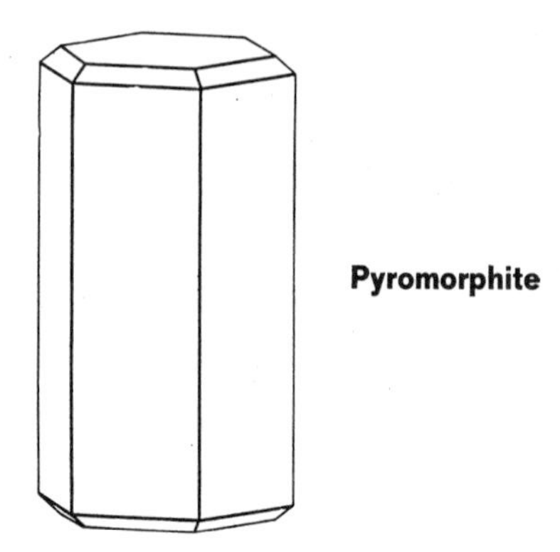

Pyromorphite

Pyromorphite. A mineral of the pyromorphite series in the apatite group. The hexagonal crystals often grow in rounded, barrel shapes which are sometimes hollow. They are heavy, resinous to adamantine, and are colored green, brown, or yellow. France, Germany, Czechoslovakia, Pennsylvania, and Idaho are among the sources. The formula is $Pb_5(PO_4,AsO_4)_3Cl$.

Pyromorphite series. A series of phosphate-arsenate-vanadate minerals in the apatite group, which have a similar origin and occurrence. The following members are included, the first two of which grade completely into each other but only partly into the third:

Pyromorphite	$Pb_5(PO_4,AsO_4)_3Cl$
Mimetite	$Pb_5(AsO_4,PO_4)_3Cl$
Vanadinite	$Pb_5(VO_4)_3Cl$

Pyrope. A subspecies of the garnet group of minerals. It is a silicate of magnesium and aluminum, $Mg_3Al_2(SiO_4)_3$. It is one of the so-called precious garnets used for deep-red gems. The principal occurrence is in peridotite and serpentine. Czechoslovakia and Arizona are sources of gem pyrope. As the variety rhodolite, pyrope grades into almandite.

Pyrophyllite. A monoclinic, silicate mineral occurring in masses. These are soft, pearly to greasy, and are colored white, green, gray, or brown. They occur in North Carolina and have industrial uses similar to talc. The formula is $Al_2Si_4O_{10}(OH)_2$.

Pyroxene cleavage

Pyroxene family. A major family of rock-forming minerals. They are silicates of various metals, of which augite has the most complex combination and is the most common. The pyroxenes may be placed in the spodumene, diopside-hedenbergite, and enstatite-hyperthene series. Or they may be classified simply as *orthopyroxenes* (when orthorhombic) and *clinopyroxenes* (when monoclinic). There is a strong resemblance between many of the pyroxene minerals and those in the related amphibole family. Pyroxene, however, has a nearly right-angled cleavage, often a good parting, and a tendency to form in stubby crystals. Pyroxene forms at a higher temperature than amphibole and is free of water.

Pyroxenoid group. A group of silicate minerals that crystallize in the triclinic system and have the same atomic structure. They are similar to the pyroxenes in composition but not in structure. The following members are included:

Wollastonite	$CaSiO_3$
Rhodonite	$MnSiO_3$
Bustamite	$CaMnSi_2O_6$
Pectolite	$Ca_2NaHSi_3O_9$

Pyrrhotite. A magnetic mineral composed of iron sulfide, $Fe_{1-x}S$. The formula is so written because pyrrhotite contains a deficiency of iron. It is one of two common minerals that are magnetic. It occurs in bronze-colored masses in Ontario (Sudbury) and in Scandinavia. The crystallization is hexagonal.

Quartz. One of the most important of minerals. The crystallization is hexagonal, and the formula is SiO_2. In its two distinct types—crystalline quartz proper and the flintlike, cryptocrystalline material known as chalcedony—it makes up a goodly proportion of the rocks of the earth's crust and appears under a wide variety of names, according to color and pattern. The tapering, striated crystals are sometimes of large size and great beauty. The luster is vitreous or greasy. The major crystalline varieties include *rock crystal, amethyst, rose quartz, milky quartz, morion, smoky quartz, cairngorm, citrine, cat's-eye quartz, tiger's-eye, sagenite,* and *aventurine.* The chalcedony varieties—which include the gems most commonly used for amateur lapidary work—include carnelian, sard, chrysoprase, plasma, agate, bloodstone, onyx, flint, chert, jasper, and prase. These are of worldwide distribution.

Quicksilver. An ancient name for mercury.

Radiated zeolite. An old name for silbite, describing the habit of this zeolite mineral.

Rainbow. A name given to several gems having an iridescent effect. Examples include *rainbow agate, rainbow quartz,* and *rainbow obsidian.*

Rasorite. Another name for kernite.

Realgar. A monoclinic mineral composed of arsenic sulfide, AsS. It has a red color and resinous luster and is almost always associated with yellow orpiment. To the collector this association is one of the most attractive of mineral combinations. Realgar is found in central Europe and in the western United States.

Red antimony. An old name for kermesite, referring to its color and chemical composition.

Red chalk. A popular name for earthy hematite, alluding to its color and texture.

Red copper ore. A miners' name for cuprite, from its color and use.

Reddingite series. A series of orthorhombic, phosphate minerals which grade into each other and have a similar origin and occurrence. It includes the following members:

Reddingite	$(Mn,Fe)_3(PO_4)_2 \cdot 3H_2O$
Phosphoferrite	$(Fe,Mn)_3(PO_4)_2 \cdot 3H_2O$

Reddle, Ruddle. Popular names for earthy hematite, referring to its red color.

Red glassy copper ore. An old name for cuprite, indicating its appearance and use.

Red iron ore. A miners' name for hematite, referring to its color and use.

Red iron vitriol. An old name for botryogen, describing its color and composition.

Red lead. An old name for minium, indicating its color and metal content.

Red ocher. An old name for earthy hematite, describing its appearance.

Red oxide of iron. An old name for hematite, indicating its color and chemical composition.

Red silver ore. An old name for pyrargyrite and proustite, from their color and use.

Red sulphuret of arsenic. An old name for realgar, indicating its color and chemical composition.

Red vitriol. An old name for bieberite, in allusion to its color and sulfate composition.

Red zink ore. An old name for zincite, because of its color and use.

Resin jack. A miners' name for sphalerite, from its resinous luster.

Rhodizite. A colorless, vitreous mineral of unusual hardness. A complex borate, $NaKLi_4Al_4Be_3B_{10}O_{27}$, it occurs in isometric crystals in Madagascar and the Soviet Union.

Rhodochrosite. A mineral of the calcite group, $MnCO_3$. It occurs in vitreous masses of pink to brown color. Various localities are in Colorado, Montana, and New Jersey. It is mined as an ore of manganese.

Rhodolite. A gem variety of garnet. It is composed of two parts of pyrope and one part of almandite. As a pale rose-red or purple gem, it comes from North Carolina.

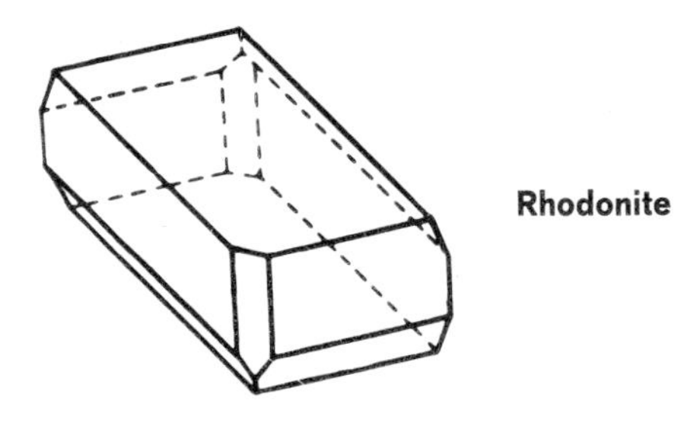

Rhodonite

Rhodonite. A pink, ornamental mineral of the pyroxenoid group. Its rose-red color is typical though not constant, and the outside may alter to black. The composition is manganese silicate, $MnSiO_3$. Some rhodonite is used as a minor ore of manganese, but the vitreous material from the Ural Mountains has long been famous for its decorative use. New South Wales (Australia), Sweden, and Brazil are other localities. The fowlerite variety comes from New Jersey.

Rhomb spar. An old name for dolomite and other similar minerals, referring to their cleavage form and bright luster on the cleavage surface.

Ribbon jasper. Jasper having broad stripes of several colors.

Ricolite. A name for verde antique (serpentine) from New Mexico.

Riebeckite. A member of the alkali-amphibole or glaucophane-riebeckite series of minerals in the amphibole family. It is often considered one of the soda amphiboles because of its sodium content.

Rock cork. An old name for asbestos and pilolite.

Rock crystal. Colorless quartz. It is the most common transparent gem. Among the most prolific sources are Brazil, Madagascar, Japan, the Alps, and Arkansas. Varieties of rock crystal include *sagenite, rainbow quartz,* and *iris quartz.*

Rock leather. Another name for mountain leather (asbestos).

Rock meal, Rock-milk. An earthy variety of calcite, named from its occurrence as soft, wet deposits in caves or on soil.

Rock ruby. An old name for red garnet, from its color.

Rock silk. An old name for asbestos.

Rock soap. An old name for several soft, slippery-feeling minerals, including montmorillonite, saponite, and oropion.

Roesslerite group. A group of arsenate minerals that crystallize in the monoclinic system, have the same atomic structure, and also grade into each other. It includes the following members:

Roesslerite	$MgH(AsO_4)\cdot 7H_2O$
Phosphorresslerite	$MgH(PO_4)\cdot 7H_2O$

Roselite group. A group of arsenate minerals that crystallize in the monoclinic system and have the same atomic structure. It includes the following members:

Roselite	$Ca_2(Co,Mg)(AsO_4)_2\cdot 2H_2O$
Brandtite	$Ca_2Mn(AsO_4)_2\cdot 2H_2O$

Rose quartz. Pink and rose-colored quartz. The cause of the color, which is often subject to fading upon exposure to light, may be titanium. South Dakota and South-West Africa are localities for material of gem quality.

Rubber-sulphur. A name given to sulfarite, suggesting its texture.

Rubellite. The pink or red, gem variety of the mineral tourmaline. It comes from Brazil, Madagascar, and California.

Rubicelle. The yellow to orange-red variety of spinel.

Ruby. A red, gem variety of the mineral corundum. It is colored by chromium oxide and grades into pink sapphire. Ruby is the most expensive gem. The best quality comes from Burma; Thailand and Ceylon are other principal sources. *Star ruby* is much less common than star sapphire. Both the clear and star varieties are made synthetically for commerce.

Ruby blende. An English and American miners' name for red sphalerite, because of its color and bright, though nonmetallic, luster.

Ruby copper. A miners' name for cuprite, from its color and metal content.

Ruby silver. Pyrargyrite or proustite. The former is called *dark ruby silver;* the latter, *light ruby silver.*

Ruby silver group. A group of sulfosalt minerals that crystallize in the hexagonal system and have the same atomic structure and similar physical properties. It includes the following members:

Pyrargyrite	Ag_3SbS_3
Proustite	Ag_3AsS_3

Ruby silver ore. An old name for pyrargyrite and proustite, indicating their color and use.

Ruby zinc. An English and American miners' name for red sphalerite, from its color and metal content.

Ruin agate. Fortification agate.

Rutile. A tetragonal mineral composed of titanium oxide, TiO_2. Its color is reddish brown and it has a bright or metallic

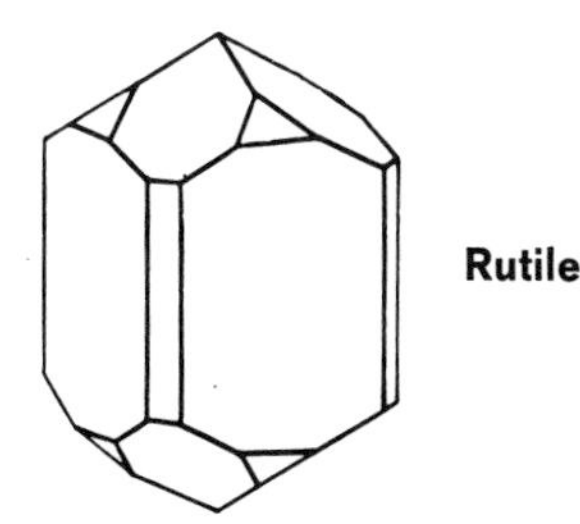

Rutile

luster. Besides occurring as crystals and in masses, rutile is common in heavy sand. Australia leads the world in the production of rutile; the United States is second. Synthetic rutile (sold as a gem called "titania") looks little like the natural mineral.

Rutile group. A group of oxide minerals that crystallize in the tetragonal system and have the same atomic structure though varying physical properties. It includes the following members:

Rutile	TiO_2
Pyrolusite	MnO_2
Cassiterite	SnO_2
Plattnerite	PbO_2

Sagenite. Rock crystal (quartz) enclosing needlelike crystals of other minerals. Among these the most familiar include rutile, tourmaline, actinolite, and goethite. Fanciful names, such as "Venus's-hairstone," "Cupid's darts," and "arrows of love," have also been used.

Saltpeter. An old, popular name for niter.

Samarskite. A radioactive mineral. It is an orthorhombic oxide of rare earths and other elements, with the formula $(Y,Er,Ce,U,Ca,Fe,Pb,Th)(Cb,Ta,Ti,Sn)_2O_6$.

Sand crystal. A crystal of calcite intermixed with quartz sand. It is shaped roughly like calcite and has a gray, sandy look with the subdued luster of calcite. South Dakota and France are well-known localities.

Sanidine. A monoclinic, high-temperature potassium feldspar, $(K,Na)AlSi_3O_8$. It occurs in volcanic rocks.

Saponite. A magnesium-bearing member of the montmorillonite group of minerals. The formula is $CaMg_3(Al,Si)_4O_{10}(OH)_2$.

Sapphire. A gem variety of the mineral corundum. The name usually indicates the blue variety, while other colors of sapphire are properly designated as pink, green, white, amethyst sapphire, etc. *Star sapphire* is a popular variety. The best blue stones come from Kashmir; Thailand, Ceylon, and Queensland (Australia) are other sources of supply. Both the clear and star varieties are made synthetically in large amounts.

Sapphirine. A blue, vitreous mineral. It is partly an oxide of magnesium, aluminum, and silicon, $(Mg,Fe)_{15}(Al,Fe)_{34}Si_7O_{80}$. It occurs in metamorphic rocks in Greenland, Madagascar, South Africa, Italy, India, and New York.

Sard. Brown chalcedony quartz. This has been a long-time favorite for small carvings. With a brighter red tone, it grades into carnelian.

Sardonyx. Onyx having bands of sard or carnelian combined with chalcedony of a different color, usually white or black. It is widely used for carving into cameos.

Satin spar. The fibrous variety of gypsum; it has a silky luster.

Sauconite. A zinc-bearing member of the montmorillonite group of clay minerals. The formula is $Ca,Na(Zn,Mg,Al,Fe)_3(Al,Si)_4O_{10}(OH)_2$.

Saussurite. A tough and compact mixture of minerals, of which zoisite is the main constituent, used as a substitute for jade. Saussurite is the result of the decomposition of feldspar. Its color is usually white, greenish, or grayish.

Scapolite. A mineral of the scapolite series, intermediate between marialite and meionite. Wernerite also represents the same composition. The occurrence is as coarse, square, tetragonal crystals which are usually vitreous and white, gray, or pale green. Important localities include Massachusetts, New York, and Ontario. A gem variety is known as *pink moonstone,* and yellow gem scapolite is found in Madagascar.

Scapolite series. A series of tetragonal, silicate minerals, which grade completely into each other. It includes the following members, as well as scapolite or wernerite, which is intermediate:

Marialite	$(Na,Ca)_4Al_3(Al,Si)_3Si_6O_{24}(Cl,CO_3,SO_4)$
Meionite	$(Ca,Na)_4Al_3(Al,Si)_3Si_6O_{24}(Cl,CO_3,SO_4)$

Scenic agate. A less common name for moss agate.

Scheelite. A tetragonal, tungstate mineral, $CaWO_4$. It occurs in heavy crystals and masses, which are vitreous to adamantine. The color is usually white, yellow, or brown, but the fluorescence is typically bluish. Nevada, California, and Australia are among the sources of scheelite.

Scheelite group. A group of tungstate-molybdate minerals that crystallize in the tetragonal system and have the same atomic structure. They grade into each other as a nearly complete series. The following members are included:

Scheelite	$CaWO_4$
Powellite	$CaMoO_4$

Schorl. The black variety of tourmaline.

Schorlomite. A garnet rich in titanium. It is a rare, black mineral.

Scolecite. A member of the zeolite family of minerals, similar to natrolite but containing calcium instead of sodium. The formula is $Ca(Al_2Si_3O_{10})\cdot 3H_2O$; the crystallization is monoclinic.

Scorodite. An orthorhombic mineral of the scorodite-mansfieldite series in the variscite group. It occurs as vitreous to adamantine crystals, mostly of green or brown color. Localities occur in Siberia, the United States, and elsewhere. The formula is $FeAsO_4\cdot 2H_2O$.

Scorodite-mansfieldite series. A series of minerals in the variscite group, which grade completely into each other. It includes the following members:

Scorodite	$Fe(AsO_4)\cdot 2H_2O$
Mansfieldite	$Al(AsO_4)\cdot 2H_2O$

Scorzalite. A blue, monoclinic mineral of the lazulite series. It occurs in vitreous masses in Brazil and South Dakota. The formula is $(Fe,Mg)Al_2(PO_4)_2(OH)_2$.

Seaweed agate. Moss agate containing fibers of the green mineral chlorite.

Selenite. The variety of gypsum occurring as colorless crystals and cleavage pieces.

Seligmannite. A mineral composed of lead-copper-arsenic sulfide, $PbCuAsS_3$. Its crystals are metallic-gray in color. Switzerland, Utah, Montana, and Western Australia are recorded as localities.

Semseyite. A monoclinic mineral composed of lead-antimony sulfide, $Pb_9Sb_8S_4$. It occurs as metallic-gray crystals in Hungary, Germany, France, Scotland, and Bolivia.

Sepiolite. A fine-textured, light-weight mineral known also as meerschaum. It is gray or white, often with a yellowish or reddish tinge. It occurs in lumps, in Turkey and elsewhere, and is used for making pipes. The formula is $Mg_4Si_6O_{15}(OH)_2\cdot 6H_2O$.

Sericite. Fine-grained muscovite mica. It has usually altered from feldspar.

Serpentine. A green, silicate mineral found in igneous and metamorphic rocks. The luster is generally greasy or waxy and the surface appearance is mottled. The usual, platy variety is called antigorite; the fibrous variety, a kind of asbestos, is chrysotile; bowenite is a gem variety. Serpentine is worldwide in distribution. The crystallization is monoclinic, and the formula is $Mg_6Si_4O_{10}(OH)_8$.

Shattuckite. A blue, copper mineral. It is opaque and occurs as granular or fibrous masses. It comes mostly from Arizona. The formula is $CuSiO_2 \cdot H_2O$.

Siberite. The reddish-violet, gem variety of tourmaline.

Sicklerite series. A series of orthorhombic, phosphate minerals, which grade into each other and have similar physical properties, as well as a similar origin and occurrence. It includes the following members:

Sicklerite	$(Li,Mn,Fe)PO_4$
Ferri-sicklerite	$(Li,Fe,Mn)PO_4$

Siderite. A mineral of the calcite group, $FeCO_3$. Its vitreous, brown crystals are usually rhombs, often curved. Siderite is an ore of iron in Great Britain, Austria, Spain, and Algeria. The preferred British name is chalybite.

Sienna earth. An old name for limonite (originally called hypoxanthite) because of its brownish-yellow color and earthy look.

Silica group. A group of minerals having the same chemical composition but different atomic structures. It includes the following members:

Quartz	SiO_2
Keatite	SiO_2
Melanophlogite	SiO_2
Tridymite	SiO_2
Cristobalite	SiO_2
Coesite	SiO_2
Stishovite	SiO_2
Lechatelierite	SiO_2
Opal	$SiO_2 \cdot nH_2O$

Siliceous sinter. A variety of opal deposited around hot springs and geysers. It is also called geyserite.

Silicified wood. Petrified wood that has been filled or replaced by chalcedony.

Sillimanite. An aluminum silicate mineral. It has the same chemical composition as andalusite and kyanite, $AlAlO(SiO_4)$, but has its own atomic structure. Its vitreous crystals are long and slender, often assuming a fibrous habit, as in the variety fibrolite. Sillimanite comes from Brazil and the United States, as well as from several countries in central Europe. A common associate of this metamorphic mineral is corundum.

Sillimanite group. A group of silicate minerals having the same chemical composition, and used in the manufacture of porcelains and refractories. It includes the following members, and sometimes dumortierite and topaz:

Sillimanite	$AlAlO(SiO_4)$
Andalusite	$AlAlO(SiO_4)$
Kyanite	$AlAlO(SiO_4)$

Silver glance. An old name for argentite because of its metal content and bright, metallic luster.

Sinhalite. A recently identified gem mineral. It is vitreous and yellow to black. The known specimens have come mainly from Ceylon. The formula is $Mg(Al,Fe)PO_4$.

Siserskite. A hard and very heavy, metallic-gray mineral. This natural alloy of osmium and iridium occurs in flakes and grains. Its localities are worldwide wherever platinum is recovered.

Sjogrenite group. A group of carbonate-hydroxide minerals that crystallize in the hexagonal system and have the same atomic structure (but different from the hydrotalcite group) and similar physical properties. It includes the following members:

Manasseite	$Mg_6Al(OH)_{16} \cdot CO_3 \cdot 4H_2O$
Barbertonite	$Mg_6Cr_2(OH)_{16} \cdot CO_3 \cdot 4H_2O$
Sjogrenite	$MgFe_2(OH)_{16} \cdot CO_3 \cdot 4H_2O$

Skutterudite. An isometric mineral composed of cobalt-nickel arsenide, $(Co,Ni)As_3$. It has a silver-gray color. It is a member of the skutterudite series.

Skutterudite series. A series of arsenide minerals that crystallize in the isometric system and have the same atomic structure and similar physical properties. It includes the following members:

Skutterudite	$(Co,Ni)As_3$
Smaltite	$(Co,Ni)As_3$
Nickel-skutterudite	$(Ni,Co)As_3$
Chloanthite	$(Ni,Co)As_3$

Smaltite. An isometric mineral composed of cobalt-nickel arsenide, $(Co,Ni)As_3$. It has a silver-gray color. It is the best known member of the skutterudite series. Smaltite yields cobalt and by-product arsenic in Ontario (Cobalt). Germany has produced fine specimens.

Smithsonite. A mineral of the calcite group, $ZnCO_3$. It usually occurs as vitreous, brown masses, but other colors are known and the mineral is often difficult to recognize. It is an ore of zinc. Dry-bone ore is a distinctive variety. Northern Rhodesia, South-West Africa, and Colorado are important localities. In Great Britain, this mineral is called *calamine.*

Smoky quartz. A smoky-yellow to dark-brown variety of quartz. It grades into cairngorm. The color is due to radioactivity. Choice specimens come from Switzerland, Spain, and Colorado (Pikes Peak region).

Sodalite. An isometric member of the feldspathoid group of minerals. Its typical occurrence is as blue masses, although crystalline occurrences and some other colors are known. Maine, Quebec, Ontario, and British Columbia (Canada) are leading sources. The formula is $Na_4(AlSiO_4)_3Cl$.

Soda niter. A hexagonal, nitrate mineral, $NaNO_3$. It has a cooling taste and usually occurs as vitreous, colorless, or white masses. The occurrences are in deserts, especially in Chile and Bolivia, where soda niter is mined as fertilizer.

Soda spar. The commercial name for albite (plagioclase) feldspar.

South African jade. A wrong name for the compact, green variety of grossularite garnet, referring to its occurrence and appearance. Another name is *Transvaal jade.*

Sparable tin. A Cornish name for cassiterite, describing the resemblance of some crystals of this tin mineral to the "sparable nail" of the shoe cobbler.

Spathic iron. An old name for siderite, because it is a kind of spar (a miners' term for nonmetallic and lustrous minerals) and contains iron.

Spear pyrites. An old name for marcasite having twinned crystals in the shape of a spear but otherwise resembling pyrite.

Specular iron. A common name for specular hematite, the chief ore of iron.

Specularite. Specular hematite.

Sperrylite. An isometric mineral composed of platinum arsenide, $PtAs_2$. It is heavy, hard, and metallic white. It is the chief ore of platinum in Ontario (Sudbury).

Spessartite. A subspecies of the garnet group of minerals. It is a silicate of manganese and aluminum, $Mn_3Al_2(SiO_4)_3$. The color is brown to red. The typical occurrence is in rhyolite, but it is found elsewhere.

Sphalerite. An important ore mineral composed of zinc sulfide, ZnS. Its properties are variable enough to cause difficulty in recognizing it. A perfect cleavage and a usually resinous luster help to identify it. Its most common color is in the yellow-to-brown range, but other colors can be expected. Sphalerite is the most important ore of zinc, cadmium, germanium, indium, and gallium. The Tri-State district (Missouri-Kansas-Oklahoma) is the world's largest source, and Canada, Mexico, the Soviet Union, and Australia are other major producers. The preferred British name is *blende*.

Sphalerite group. A group of sulfide-telluride minerals that crystallize in the isometric system and have the same atomic structure and similar physical properties, except that sphalerite does not have a metallic luster. The following members are included:

Sphalerite	(Zn,Fe)S
Metacinnabar	(Hg,Fe,Zn)S
Tiemannite	HgSe
Coloradoite	HgTe

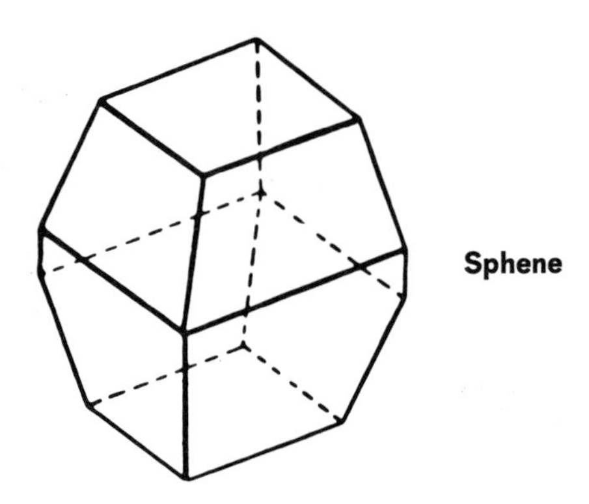
Sphene

Sphene. An industrial and, although rather soft (5 to 5½), an occasional gem mineral. It is a silicate of calcium and titanium, $CaTiO(SiO_4)$. The monoclinic crystals are wedge shaped and have a bright luster; its color is usually brown. Specimens of good quality come from Norway, Switzerland, Italy, Canada, and the United States. Sphene is an ore of titanium and is mined on the Kola Peninsula of the Soviet Union.

Spinel. A fine gem mineral, resembling ruby *(ruby spinel)*, sapphire, and other gems. *Balas ruby* is rose colored; *rubicelle* is yellow to orange-red; *almandine-spinel* is violet; *picotite* is yellowish to greenish brown to blackish; *pleonaste* is dark green to black. Rounded, hard (8), gemmy crystals come from placer sand in Madagascar, Ceylon, Burma, and Thailand. Common spinel is found in New York and New Jersey. The crystallization is isometric, and the composition is magnesium-aluminum oxide, $MgAl_2O_4$.

Spinel group. A group of oxide minerals that crystallize in the isometric system and have the same atomic structure and similar physical properties. It includes the following members:

Spinel series
Magnetite series
Chromite series

Spinel series. A series of oxide minerals that crystallize in the isometric system and have the same atomic structure and similar physical properties, according to chemical composition. It belongs to the spinel group and includes the following members:

Spinel	$MgAl_2O_4$
Hercynite	$FeAl_2O_4$
Gahnite	$ZnAl_2O_4$
Galaxite	$MnAl_2O_4$

Spodumene. A mineral of the spodumene series of minerals in the pyroxene family. It is noted for its huge crystals which occur in pegmatite in sizes as heavy as 90 tons. Nevertheless, spodumene is not a common mineral. It is a silicate of lithium and aluminum, $LiAl(Si_2O_6)$. Its usual color is light gray, but occasional gem varieties are green *(hiddenite)* and lilac *(kunzite)*. The luster is vitreous. As a source of lithium and for use in ceramics, spodumene comes mostly from North Carolina and South Dakota.

Spodumene series. A series of monoclinic, silicate minerals in the pyroxene family. It includes the following members:

Spodumene	$LiAl(Si_2O_6)$
Jadeite	$NaAl(Si_2O_6)$
Aegirite	$NaFe(Si_2O_6)$

Stannite. A tetragonal mineral composed of copper-tin-iron sulfide, Cu_2FeSnS_4. It is the only ore of tin except cassiterite and is mined as metallic-gray masses in Bolivia.

Star stone. A variety of sapphire, ruby, garnet, quartz, and certain other minerals showing asterism.

Staurolite. A mineral noted for its brown, hard (7 to 7½), cross-shaped, orthorhombic, twin crystals. These form in several kinds of crosses, the most familiar being the *fairy stones* used as ornaments and amulets. Staurolite is a silicate of iron and aluminum, $Fe_2Al_9O_7(SiO_4)_4(OH)$. Some of the best localities are in

France, Scotland, Switzerland, Czechoslovakia, Germany, and the United States.

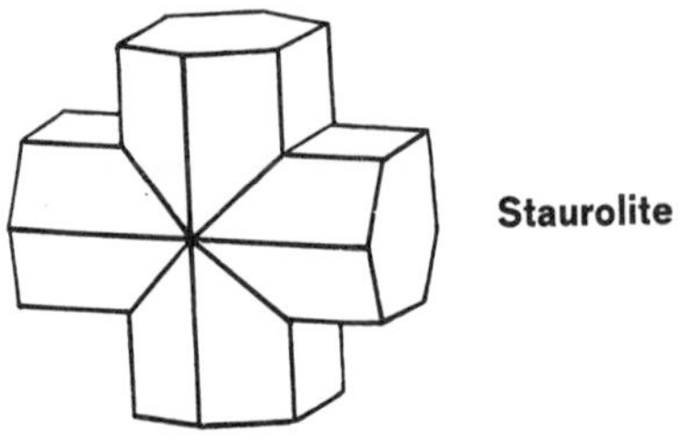

Staurolite

Stephanite. An orthorhombic mineral composed of silver-antimony sulfide, Ag_5SbS_4. It has the color of iron and is brittle though soft. In addition to its important occurrence in Nevada (Comstock Lode), it is found in other silver-mining districts of the world.

Stibiopalladinite. A heavy mineral composed of palladium antimonide, Pd_3Sb. It occurs in metallic-gray fragments in South Africa.

Stibiotantalite series. A series of oxide minerals that crystallize in the orthorhombic system and have the same atomic structure and similar physical properties, according to the chemical composition. It includes the following members:

Stibiotantalite	$SbTaO_4$
Stibiocolumbite	$SbCbO_4$

Stibnite. An attractive mineral of the stibnite group, composed of antimony sulfide, Sb_2S_3. It is lead gray in color and has a perfect cleavage. Its orthorhombic crystals often look bent and show lines or grooves. The crystals from Japan are outstanding among mineral specimens. China and Mexico are large producers of stibnite.

Stibnite group. A group of sulfide-selenide minerals that crystallize in the orthorhombic system and have the same atomic structure and physical properties, although those of guanajuatite are little known. The following members are included:

Stibnite	Sb_2S_3
Bismuthinite	Bi_2S_3
Guanajuatite	Bi_2Se_3

Stichtite. A rose-colored to lilac mineral. It is a carbonate-hydroxide of magnesium and chromium, $Mg_6Cr_2(OH)_{16}CO_3 \cdot 4H_2O$. It occurs in Tasmania (Australia), South Africa, Algeria, and Canada.

Stilbite. A member of the stilbite group of minerals in the zeolite family. It often occurs in sheaflike groups of monoclinic crystals, which are white or tinged from yellow to red. The cleavage is prominent, giving a pearly luster on the side face. The formula is $Ca(Al_2Si_7O_{18})\cdot 7H_2O$. The preferred British name is *desmine.*

Stilbite group. A group of minerals in the zeolite family. It includes the following members:

Stilbite	$Ca(Al_2Si_7O_{18})\cdot 7H_2O$
Phillipsite	$KCa(Al_3Si_5O_{16})\cdot 6H_2O$
Harmotome	$Ba(Al_2Si_6O_{16})\cdot 6H_2O$

Stream tin. A miners' name for pebbles of cassiterite (a tin ore) found in placers.

Strengite. A member of the variscite-strengite series of minerals in the variscite group. It occurs as rounded, fibrous aggregates and as crusts; the color is pink or violet and the luster is vitreous. The localities are in Germany, Portugal, Sweden, and elsewhere. The formula is $FePO_4\cdot 2H_2O$, and the crystallization is orthorhombic.

Stromeyerite. An orthorhombic mineral composed of silver-copper sulfide, $(Ag,Cu)_2S$. This metallic-gray mineral has been found in Siberia, Tasmania (Australia), and a number of places in the Western Hemisphere.

Strontianite. A mineral of the aragonite group; its formula is $SrCO_3$. It is vitreous and usually white. It has a number of minor commercial uses. Germany, Spain, Mexico, and England have some of the main deposits.

Sulfur group. A group of mineral elements that includes three forms of native sulfur. Of these, *alpha-sulfur* is the familiar one always referred to when native sulfur is mentioned.

Sulphuret. An old name applied to many minerals, mostly metallic, that contain sulfur in addition to other elements. Among these are such familiar ones as *sulphuret of antimony* (stibnite), *of copper* (chalcocite), *of silver* (argentite), *of zinc* (sphalerite), and such interesting ones as *triple sulphuret of lead, antimony, and copper* (bournonite), *red sulphuret of arsenic* (realgar), and *magnetic sulphuret of iron* (pyrrhotite).

Sulvanite group. A group of sulfosalt minerals that crystallize in the isometric system and are somewhat related in chemical composition and physical properties. It includes the following members:

Sulvanite	Cu_3VS_4
Germanite	$(Cu,Ge,Fe,Zn,Ga)(S,As)$
Colusite	$Cu_3(As,Sn,V,Fe,Te)S_4$

Sunstone. A gem variety of oligoclase (plagioclase) feldspar. A sparkling effect is given by hematite inclusions. Norway is the principal source.

Supersulphuretted lead. An old name for galena, because of its chemical composition.

Sussexite series. A series of orthorhombic, borate minerals, which grade into each other more or less completely. It includes the following members:

Sussexite	$(Mn,Mg)(BO_2)(OH)$
Szaibelyite	$(Mg,Mn)(BO_2)(OH)$

Svabite series. A series of hexagonal, arsenate minerals in the apatite group, which grade into each other. It includes the following members:

Svabite	$(Ca,Pb)_5(AsO_4,PO_4)_3(F,Cl,OH)$
Hedyphane	$(Ca,Pb)_5(AsO_4)_3Cl$

Swiss jade. A wrong name for green-dyed jasper, because of its appearance.

Swiss lapis. A wrong name for blue-dyed jasper, because of its appearance.

Sylvanite. A heavy, monoclinic mineral composed of gold-silver telluride, $(Au,Ag)Te_2$. It is silver colored, but yields a mixed button of gold and silver when the tellurium is driven off by heat. Colorado, California, Oregon, Ontario, Rumania, and Western Australia are the main sources of this ore of gold and silver.

Sylvite. An isometric mineral composed of potassium chloride, KCl. It occurs mostly in granular, cleavable masses that have a bitter taste. New Mexico, Texas, Germany, and Poland are leading sources of sylvite, which often occurs with halite as a product of evaporation.

Taafeite. A recently discovered gem mineral. It is violet, transparent, and vitreous. It is an aluminate of beryllium and magnesium, $BeMgAl_4O_8$, and is believed to come from Ceylon.

Tabasheer. A variety of opal of organic origin. It is found inside bamboo.

Tabular spar. An old name for wollastonite, referring to its crystal form and its bright luster on the cleavage surface.

Talc. The softest known mineral. It occurs in pearly to greasy masses which are green, gray, or white and have a slippery feel. Talc is found in altered and metamorphic rocks and is produced for various industrial uses in the Appalachians, Pyrenees, and other mountain ranges. The crystallization is monoclinic, and the formula is $Mg_3Si_4O_{10}(OH)_2$.

Tamarugite group. A group of sulfate minerals that crystallize in the monoclinic system and have the same atomic structure and a similar occurrence. It includes the following members:

Tamarugite	$NaAl(SO_4)_2 \cdot 6H_2O$
Amarillite	$NaFe(SO_4)_2 \cdot 6H_2O$

Tantalite. An orthorhombic mineral composed of tantalum-columbium oxide, $(Fe,Mn)(Ta,Cb)_2O_6$. It occurs in heavy, iron-black crystals. It is a member of the columbite-tantalite series and is an ore of both of its rare metals. Tantalite comes from pegmatites and placers in the United States, Canada, and Sweden.

Tapiolite series. A series of oxide minerals that crystallize in the tetragonal system and have the same atomic structure and similar physical properties, according to the chemical composition. It includes the following members:

Tapiolite	$FeTa_2O_6$
Mossite	$Fe(Cb,Ta)_2O_6$

Tellurium group. A group of native-element minerals that crystallize in the hexagonal system and have the same atomic structure. It includes the following members:

Native selenium	Se
Native tellurium	Te

Tellurobismuthite. A soft, heavy mineral having a metallic luster and gray color. It is a member of the tetradymite group and has the formula Bi_2Te_3, crystallizing in the hexagonal system. It has come from Japan, Sweden, Canada, and scattered localities in the United States, especially Colorado.

Tennantite. An isometric mineral composed of copper-arsenic sulfide, $Cu_{12}As_4S_{13}$. It is a member of the tetrahedrite series and is found in the same places as tetrahedrite but less commonly.

Tenorite. A triclinic mineral composed of copper oxide, CuO. A metallic-black mineral, it occurs in masses known as melaconite. This material is found in copper mines everywhere.

Tephroite. A manganese silicate mineral found with rhodonite. Its color is red to gray. The crystallization is orthorhombic, and the formula is $Mn_2(SiO_4)$.

Tetradymite. A mineral occurring in heavy, soft, metallic-gray masses. It is a member of the tetradymite group and is found in numerous deposits around the world. The crystallization is hexagonal, and the formula is Bi_2Te_2S.

Tetradymite group. A group of telluride-sulfide minerals that crystallize in the hexagonal system and have a similar atomic structure and physical properties. It includes the following members:

Tellurobismuthite	Bi_2Te_3
Tetradymite	Bi_2Te_2S
Gruenlingite	Bi_4TeS_3

Tetrahedrite. The most abundant sulfosalt. It is an isometric mineral composed of copper-antimony sulfide, $Cu_{12}Sb_4S_{13}$. As a member of the tetrahedrite series, it grades into tennantite. The color is metallic gray, often bright. Tetrahedrite serves as an ore of copper, silver (as the variety *freibergite),* and antimony. The western part of the United States and Canada has produced much tetrahedrite, as has South-West Africa.

Tetrahedrite series. A series of sulfosalt minerals that crystallize in the isometric system and have the same atomic structure and similar physical properties. It includes the following members:

Tetrahedrite	$(Cu,Fe)_{12}Sb_4S_{13}$
Tennantite	$(Cu,Fe)_{12}As_4S_{13}$

Thetis' hair-stone. A variety of quartz containing needlelike inclusions of actinolite.

Thomsonite. A member of the zeolite group of minerals. It often shows an eye pattern. The formula is $(Ca,Na)_6Al_8(Al,Si)_2Si_{10}O_{40} \cdot 12H_2O$.

Thorianite. A radioactive, isometric mineral composed of thorium oxide, ThO_2. It is hard and heavy, black, and almost metallic looking. Ceylon and Madagascar are leading localities.

Thorite. A radioactive mineral belonging to the zircon group. It is a thorium silicate, $ThSiO_4$, usually occurring in irregular masses and grains of moderate hardness. It varies in density according to the degree of radioactive alteration and chemical change. A resinous luster is typical, and the color may be black or reddish brown; orangite is the orange-colored variety. Uranothorite is a variety containing uranium; and the variety called thorogummite has hydroxyl (OH) replacing the silicate ion.

Thorogummite. The hydroxyl-bearing variety of the mineral thorite.

Thulite. The pink variety of the mineral zoisite, colored by manganese.

Thunder egg. Nodules of chalcedony filled with angular or banded patterns of color. These often show "scenes" of water or landscape.

Tiemannite. A mineral composed of mercury selenide, HgSe. It is heavy, metallic gray in color, and is found mostly in masses. Germany and Utah contain the main deposits.

Tiger's-eye. A gem variety of quartz. Its rich, golden-brown, wavy bands of light make it popular for cameos and other cut stones. This is a pseudomorph of quartz after crocidolite, the original blue asbestos of which has been oxidized brown and replaced by silica. South Africa (Griqualand West) is the source.

Tile ore. An old name for earthy cuprite, because when mixed with clay it is used for making tiles.

Tinder-ore. An old name of uncertain meaning, used for a mixture of impure jamesonite and other minerals.

Tin pyrites. An old name for stannite because of its content of tin and its chemical composition (similar to pyrite).

Tin stone. A miners' name for cassiterite, because of its metal content.

Tin white cobalt. An old name for smaltite, referring to its color and metal content.

Titania. Trade name for a synthetic rutile gem. This material, being a very pale yellow, does not look like the natural mineral.

Titanite. A less acceptable name for the mineral sphene, referring to its chemical composition.

Toad's-eye tin. A common name for rounded pieces of cassiterite (the chief tin mineral) showing a concentric pattern.

Topaz. An outstanding gem mineral. It is an aluminum silicate containing fluorine, $Al_2(SiO_4)(F,OH)_2$. Its hard (8), orthorhombic crystals are distinctively shaped and are identified by their cleavage, vitreous luster, and high density. Sometimes they weigh hundreds of pounds. Although topaz shows a range of colors, the hues are readily recognized; pink is usually artificially created by heat treatment. Gem topaz comes especially from Brazil, the Soviet Union, Germany, Nigeria, Japan, Mexico, and the United States. Most topaz sold in jewelry is actually yellow quartz. Massive topaz, such as that from South Carolina, is used industrially as a refractory. Topaz originates in igneous rocks, especially pegmatite, and in quartz veins.

Topazolite. A yellow variety of andradite garnet.

Torbernite group. A group of phosphate-arsenate minerals that crystallize in the tetragonal system and have the same atomic structure and similar physical properties, especially radioactivity. The following members are included:

Torbernite	$Cu(UO_2)_2(PO_4)_2 \cdot 8\text{–}12H_2O$
Autunite	$Ca(UO_2)_2(PO_4)_2 \cdot 10\text{–}12H_2O$
Uranocircite	$Ba(UO_2)_2(PO_4)_2 \cdot 8H_2O$
Saleeite	$Mg(UO_2)_2(PO_4)_2 \cdot 10H_2O$
Zeunerite	$Cu(UO_2)_2(AsO_4)_2 \cdot 10\text{–}16H_2O$
Uranospinite	$Ca(UO_2)_2(AsO_4)_2 \cdot 8H_2O$

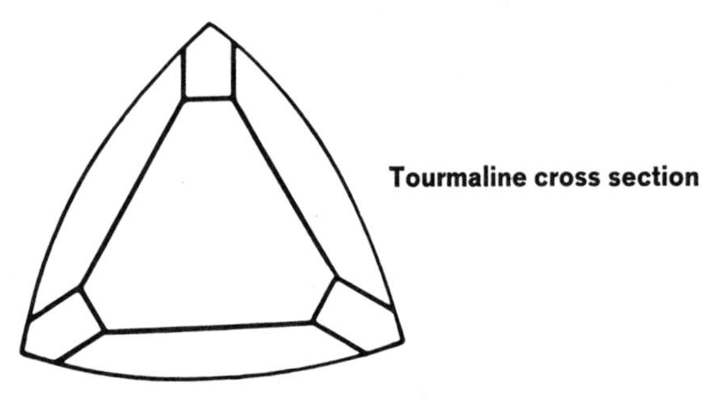
Tourmaline cross section

Tourmaline. A gem mineral of remarkable interest. It occurs in a wide range of colors, in prismatic, vitreous crystals. These are hard (7 to 7½), typically have a triangular outline, are striated vertically, and show color zones vertically or horizontally. Tourmaline is an extremely complex silicate, containing boron. According to their color, which is mainly pink, green, and blue, the gem varieties go under such names as *rubellite* and *indicolite.* Most tourmaline, however, is black and is called *schorl;* the variety called *dravite* is brown; colorless tourmaline is called *achroite.* The noted localities include Brazil, the Ural Mountains, Elba, Madagascar, and the United States (especially California, New York, and New England). Owing to its piezoelectricity, tourmaline is used in pressure gauges, such as those used in submarines.

Transvaal jade. A wrong name for the compact, green variety of grossularite garnet, referring to its occurrence and appearance. Another name is *South African jade.*

Travertine. Calcite deposited by hot- or cold-spring water. It is more or less porous. In some places it is quarried as a building stone.

Tree agate. Moss agate showing a treelike pattern.

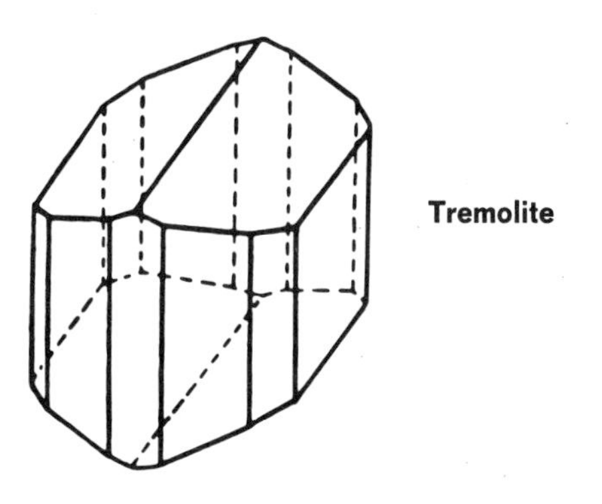

Tremolite

Tremolite. A member of the tremolite-actinolite series of minerals in the amphibole family. It is a hydrous silicate, $Ca_2Mg_5Si_8O_{22}(OH)_2$, and is white and vitreous. With a substitution of iron for magnesium, tremolite becomes green and thereby grades into actinolite. When compact and tough, either mineral is called *nephrite,* which is one of the true jade minerals. Tremolite crystals of typical monoclinic, prismatic shape are found in Switzerland, Austria, Italy, and New York.

Tremolite-actinolite series. A series of monoclinic, silicate minerals in the amphibole family, which grade completely into each other. It includes the following members:

Tremolite	$Ca_2Mg_5(Si_8O_{22})(OH)_2$
Actinolite	$Ca_2(Mg,Fe)_5(Si_8O_{22})(OH)_2$

Tridymite. A mineral of the silica group, SiO_2. It occurs as twinned, orthorhombic crystals in volcanic rock. Colorado is a noted source.

Tripe stone. An old name for anhydrite, referring to its occasional appearance.

Triphylite. An orthorhombic, phosphate mineral of the triphylite-lithiophilite series. It occurs in vitreous to resinous masses that are gray or stained almost black. They are found in pegmatite, especially in Germany, Finland, and New Hampshire. The formula is $LiFePO_4$.

Triphylite group. A group of phosphate minerals that crystallize in the orthorhombic system and have the same atomic structure. It includes the following members:

Triphylite-lithiophilite series
Hühnerkobelite-varulite series

Triphylite-lithiophilite series. A series of phosphate minerals in the triphylite group, which grade completely into each other and have similar physical properties, as well as a similar origin and occurrence. It includes the following members:

Triphylite	$Li(Fe,Mn)PO_4$
Lithiophilite	$Li(Mn,Fe)PO_4$

Triple sulphuret of lead, antimony, and copper. An old name for bournonite, from its chemical composition.

Triploidite group. A group of phosphate minerals that crystallize in the monoclinic system and have the same atomic structure, grading into each other as a complete series. They have a similar origin and occurrence. The following members are included, as well as perhaps the arsenate mineral called sarkinite:

Triploidite	$(Mn,Fe)_2(PO_4)(OH)$
Wolfeite	$(Fe,Mn)_2(PO_4)(OH)$

Trona. A monoclinic, carbonate mineral, $Na_3H(CO_3)_2 \cdot 2H_2O$. It is an evaporite mineral found in many deserts.

Troostite. A manganese variety of the mineral willemite. It occurs in New Jersey (Franklin area) in large, reddish crystals.

Tungstic ocher. A common name for tungstite, referring to its metal content and earthy look.

Turgite. A mineral composed of hydrous iron oxide, $2Fe_2O_3 \cdot H_2O$. The earthy and rounded masses are red.

Turkey-fat ore. A miners' name for cadmium-bearing, yellow smithsonite (a zinc ore), alluding to its appearance.

Turquoise. A gem mineral of the turquoise group. It occurs as nodules and veins having a waxy luster and a color in the range of blue and green. The main occurrences are in Iran and the southwestern part of the United States. The crystallization is triclinic, and the formula is $CuAl_6(PO_4)_4(OH)_8 \cdot 4H_2O$.

Turquoise group. A group of phosphate minerals, perhaps grading completely into each other, that crystallize in the triclinic system and have the same atomic structure. It includes the following members:

Turquoise	$CuAl_6(PO_4)_4(OH)_8 \cdot 4H_2O$
Chalcosiderite	$CuFe_6(PO_4)_4(OH)_8 \cdot 4H_2O$

Tyuyamunite. A radioactive mineral. It occurs as a yellow, earthy powder, which fluoresces yellow-green. It is found in the Soviet Union and the western United States. The formula is $Ca(UO_2)_2(VO_4)_2 \cdot nH_2O$.

Ulexite. A triclinic, borate mineral. It is also known as *cotton-ball borax* because it occurs as loose, rounded masses consisting of fine, white, silky fibers. Sources include Chile, Argentina, Nevada, and California. The formula is $NaCaB_5O_9 \cdot 8H_2O$. Mineral collectors are intrigued by the curious optical effect imposed by a cut plate of ulexite, which appears to raise the image of an object the thickness of the superimposed plate.

Ullmannite. An isometric mineral composed of nickel sulfantimonide, NiSbS. It occurs in metallic-gray crystals. Sardinia is an outstanding locality.

Umangite. A mineral composed of copper selenide, Cu_3Se_2. Its metallic-red masses tarnish easily to an iridescent blue. Argentina, Germany, and Sweden are among its sources.

Uraninite. The chief mineral source of radioactivity. It is an isometric mineral composed of uranium oxide, UO_2. When impure, it is known as *pitchblende.* Both varieties of this heavy, black substance are strongly radioactive and are the principal source of uranium, radium, and nuclear energy. The main localities are in the United States, Canada, Czechoslovakia, the Congo, and South Africa.

Uraninite group. A group of oxide minerals that crystallize in the isometric system and have the same atomic structure and related physical properties. It includes the following members:

Uraninite	UO_2
Thorianite	ThO_2

Uran-mica. An old name for torbernite, indicating its chemical composition and micaceous cleavage.

Uranochre. An old name for several uranium-bearing minerals, including uraconite, uranopilite, and others of uncertain identity.

Uranothorite. The uranium-bearing variety of the mineral thorite.

Uran-vitriol. An old name for johannite, indicating its metal and sulfate composition.

Uvarovite. A subspecies of the garnet group of minerals. It is a silicate of calcium and chromium, $Ca_3Cr_2(SiO_4)_3$. The color is green. Associated with chromite in serpentine, uvarovite is seldom large enough to be cut as a gem.

Vanadic ocher. An old name for cuprite, based on a wrong analysis. The name has also been applied to other minerals.

Vanadinite. A member of the pyromorphite series of minerals, in the apatite group. The hexagonal crystals are heavy and have a red, brown, or yellow color and an adamantine to resinous luster. *Endlichite* is an arsenic-bearing variety. The best localities are in South-West Africa, Morocco, Arizona, and New Mexico. It is an ore of vanadium and lead, the formula being $Pb_5Cl(VO_4)_3$.

Variegated copper ore. An old name for bornite, referring to its iridescent tarnish and its use.

Variscite. A gem mineral of the variscite-strengite series of the variscite group. It is bluish green and is found in nodules in Utah. The formula is $AlPO_4 \cdot 2H_2O$, and the crystallization is orthorhombic.

Variscite group. A group of phosphate-arsenate minerals that crystallize in the orthorhombic system and have the same atomic structure. It includes the following members:

Variscite-strengite series
Scorodite-mansfieldite series

Variscite-strengite series. A series of orthorhombic, phosphate minerals in the variscite group which grade completely into each other. It includes the following members:

Variscite	$Al(PO_4) \cdot 2H_2O$
Strengite	$Fe(PO_4) \cdot 2H_2O$

Velvet copper ore. An old name for cyanotrichite, because of its appearance and use.

Venus's-hairstone. A popular name for rock crystal (quartz) enclosing needlelike crystals of other minerals. The mineralogical name is *sagenite*.

Vermiculite. A micaceous mineral that expands when heated. It therefore makes a good insulator and lightweight building material. Vermiculite seems to be an alteration product of biotite and phlogopite mica.

Vesuvian garnet. An old name for leucite, from its occurrence at Mount Vesuvius and its crystal form which is the same as garnet.

Vesuvianite. A less acceptable name for the mineral idocrase. It was named for its original discovery at Mount Vesuvius.

Violane. Another name for diopside.

Viridine. The green variety of the mineral andalusite.

Visor tin. A miners' name for the twin crystals of cassiterite.

Vitreous copper. An old name for chalcocite, referring to its bright luster and metal content.

Vitreous silver. An old name for argentite, referring to its bright luster and metal content.

Vitriol. A name used to indicate a sulfate mineral, such as *green vitriol* (melanterite), *cobalt vitriol* (bieberite), *blue vitriol* (chalcanthite), and *zinc vitriol* (goslarite).

Vitriol ocher. An old name for glockerite, indicating its sulfate content and earthy look.

Vivianite. A monoclinic, phosphate mineral, a member of the vivianite group. The crystals readily alter from colorless to blue or green and are vitreous or pearly. Bolivia, Japan, New Jersey, and Colorado are major sources. The formula is $Fe_3(PO_4)\cdot 8H_2O$.

Vivianite group. A group of phosphate-arsenate minerals that crystallize in the monoclinic system and have the same atomic structure. It includes the following members:

Vivianite	$Fe_3(PO_4)_2\cdot 8H_2O$
Erythrite-annabergite series	
Koettigite	$Zn_3(AsO_4)_2\cdot 8H_2O$

Wad. A mixture of manganese oxides. This field term is used to include pyrolusite, psilomelane, and other minerals that are difficult to tell apart.

Water sapphire. A wrong name for the mineral iolite, referring to its appearance.

Wavellite. An orthorhombic, phosphate mineral. It appears as rounded aggregates having a radial structure. The luster is vitreous, and the color is green, brown, yellow, or white. Bolivia, Pennsylvania, and Arkansas are leading sources.

Wernerite. A mineral of the scapolite series, intermediate between marialite and meionite. Scapolite proper also represents the same composition.

Wheel-ore. An English miners' name for bournonite, describing its twinned crystals.

Whewellite. A mineral composed of calcium oxalate, $CaC_2O_4 \cdot H_2O$. It is thus of organic nature and is usually of organic origin. It comes from a number of localities in Europe.

White antimony. An old name for valentinite and cervantite, indicating their common color and chemical composition.

White arsenic. An old name for arsenolite, indicating its color and chemical composition.

White copper. A miners' name for domeykite, because of its color and metal content.

White copperas. An old name for goslarite and coquimbite, because of their color and sulfate composition.

White garnet. A popular name for leucite, alluding to its color and garnetlike crystal form.

White iron ore. An old name for light-colored siderite, because of its color and use.

White iron pyrites. A miners' name for marcasite, because of its paler color compared to pyrite.

White lead ore. An old name for cerussite, because of its color and use.

White mica. Muscovite, so called from its usual color.

White nickel. An old name for rammelsbergite and chloanthite, because of their color and metal content.

White opal. Precious opal having a light-colored background; this is usually white but may be yellow or pink. It is also known as *Hungarian opal* because of its best-known source, now in Czechoslovakia. New South Wales (Australia) and South Australia are leading producers. *Harlequin opal* and *lechosos opal* are special varieties of white opal.

White pyrites. An old name for marcasite, indicating its lighter color compared to pyrite.

White tellurium. An old name for sylvanite, indicating its usual color and chemical composition.

White vitriol. An old name for goslarite, referring to its color and sulfate composition.

Willemite. A zinc silicate mineral, $ZnSiO_4$. Although found in Belgium, Algeria, the Congo, Northern Rhodesia, South-West Africa, Greenland, New Mexico, and Arizona, the dominant locality is in New Jersey (Franklin area). Here it is closely associated with zincite and franklinite and is mined for zinc. The willemite is yellow green or reddish brown, usually in vitreous masses, and it fluoresces and phosphoresces a vivid green. Troostite is a manganese-bearing variety of willemite.

Witherite. A mineral of the aragonite group, $BaCO_3$. It occurs mostly as white or gray crystals. The principal sources are in England and the United States.

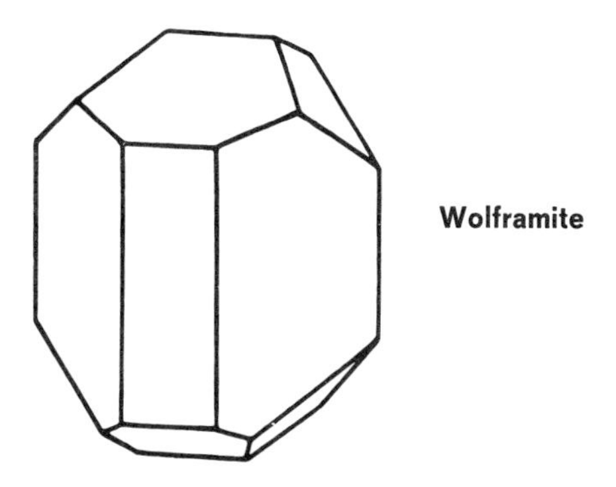

Wolframite

Wolframite. A member of the wolframite series of minerals. It is dark brown and has a submetallic to resinous luster. The usual material is in bladed forms of considerable density. China, Burma, Bolivia, and New South Wales (Australia) are leading sources of this most important ore of tungsten. The formula is $(Fe,Mn)WO_4$, and the crystallization is monoclinic.

Wolframite series. A series of monoclinic, tungstate minerals which grade completely into one another and have related physical properties. It includes the following members:

Wolframite	$(Fe,Mn)WO_4$
Huebnerite	$MnWO_4$
Ferberite	$FeWO_4$

Wollastonite. A white, industrial mineral belonging to the pyroxenoid group. Sometimes it is fibrous and may appear silky, pearly, or vitreous, according to its structure. Wollastonite is calcium silicate, $CaSiO_3$, and is used in making ceramics, paint, and insulation. Large bodies of it are known in New York, California, France, and Mexico.

Wood-copper. An old name for fibrous olivenite, owing to its appearance and metal content.

Wood opal. Another name for opalized wood.

Wood-tin. A common name for rounded pieces of cassiterite, the main tin mineral, showing a concentric pattern.

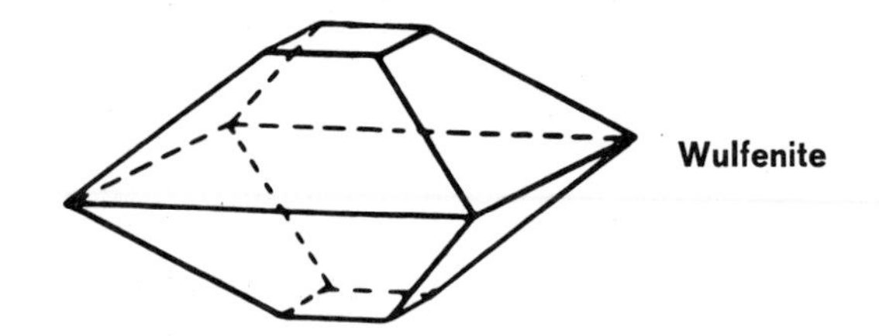

Wulfenite. A tetragonal, molybdate mineral. It occurs in bright, tabular crystals usually having a yellow, orange, or red color. It is found with other lead-bearing minerals in Arizona, Nevada, Utah, and New Mexico. The formula is $PbMoO_4$.

Wulfenite group. A group of molybdate-tungstate minerals that crystallize in the tetragonal system and have the same atomic structure. They grade into each other and form at least a partial series. The following members are included:

Wulfenite	$PbMoO_4$
Stolzite	$PbWO_4$

Wurtzite. A mineral composed of zinc sulfide, ZnS. It has the same composition as sphalerite and also similar properties, but it is hexagonal instead of isometric.

Wurtzite group. A group of sulfide minerals that crystallize in the hexagonal system and have the same atomic structure and similar physical properties, although the oxysulfide voltzite is inadequately known. The following members are included:

Wurtzite	ZnS
Greenockite	CdS
Voltzite	Zn_5S_4O

Xanthophyllite. A member of the brittle-mica group of minerals. Its formula is $CaMg_3(Al_2Si_2O_{10})(OH)_2$.

Yellow arsenic. An old name for orpiment, from its color and chemical composition.

Yellow copper ore. An old miners' name for chalcopyrite, indicating its color and use.

Yellow lead ore. An old miners' name for wulfenite, referring to its color and use.

Yellow ocher. An old name for limonite, describing its color and earthy look.

Yellow sulphuret of arsenic. An old name for orpiment, owing to its color and composition.

Yellow tellurium. An old name for sylvanite, indicating its occasional yellow tinge and its chemical composition.

Zeolite family. An interesting family of silicate minerals having somewhat related atomic structures and chemical compositions, as well as similar physical properties, especially dehydration and ion exchange. They appear to boil when heated, giving off water continuously. In turn, they absorb various liquids and gases. They also serve to soften water by exchanging their "soft" sodium for the "hard" calcium in the water. The zeolites occur as late minerals in the rock-forming process, principally in cavities and veins in lava rocks. The main sources are such famous localities as Nova Scotia, New Jersey, Iceland, India, and Italy. Associated with them are calcite, apophyllite, prehnite, pectolite, and datolite. The following members are the best known zeolites:

Analcime	$Na(AlSi_2O_6)H_2O$
Natrolite	$Na_2(Al_2Si_3O_{10})\cdot 2H_2O$
Scolecite	$Ca(Al_2Si_3O_{10})\cdot 3H_2O$
Chabazite	$(Ca,Na)_2(Al_2Si_4O_{12})\cdot 6H_2O$
Laumontite	$Ca(Al_2Si_4O_{12})\cdot 4H_2O$
Heulandite	$Ca(Al_2Si_7O_{18})\cdot 6H_2O$
Stilbite	$Ca(Al_2Si_7O_{18})\cdot 7H_2O$

Zinc blende. A miners' name for sphalerite, from its deceiving, or "blind" (nonproductive), resemblance to galena.

Zinc bloom. An old name for hydrozincite occurring as crusts.

Zincite. A hexagonal mineral composed of zinc oxide, ZnO. It forms the red material in the famous association with franklinite, willemite, and calcite from New Jersey (Franklin and Sterling Hill). Zincite is rare elsewhere.

Zincite group. A group of oxide minerals that crystallize in the hexagonal system and have the same atomic structure but differ appreciably in hardness and specific gravity. It includes the following members:

Zincite	ZnO
Bromellite	BeO

Zinc-spar. An old name for smithsonite, referring to its metal content and bright luster on the cleavage surface.

Zinc vitriol. An old name for goslarite, indicating its metal content and sulfate composition.

Zinkenite. A hexagonal mineral composed of lead-antimony sulfide, $Pb_6Sb_{14}S_{27}$. It has a metallic-gray color and a fibrous look, being one of the feather ores. Germany is perhaps the best-known locality.

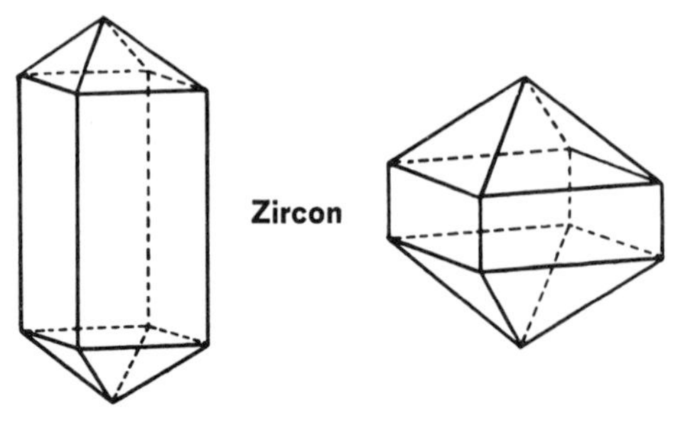

Zircon. A gem and industrial mineral of great interest. It is a zirconium silicate, $SrSiO_4$, and a member of the zircon group, occurring mostly in lustrous, brown crystals. These are easily recognized from their simple, tetragonal forms of prisms and pyramids; they are heavy and hard (7½). The radioactivity of associated thorium or uranium may break down the crystalline structure. Zircon occurs in igneous rocks and in placers. The clear varieties used as gems include hyacinth, jacinth, and jargon, but most commercial gem zircon is heat treated to yield the popular blue, golden, and colorless stones. Ceylon and Australia contain important localities for gem material, whereas good crystals of ordinary zircon come from Madagascar, Ontario, and North Carolina. The mineral is an ore of zirconium and hafnium; vast sand deposits in Queensland (Australia) and Florida are the chief sources.

Zircon group. A group of silicate minerals that crystallize in the tetragonal system and have the same atomic structure. It includes the following members:

Zircon	$Zr(SiO_4)$
Thorite	$Th(SiO_4)$

Zoisite. A member of the epidote group of minerals. It is a hydrous silicate of calcium and aluminum, $Ca_2Al_3Si_3O_{12}(OH)$. The long, prismatic, vitreous crystals are usually gray. The pink variety is called *thulite*.

ROCKS

A rock is a natural body of material of which the earth is made. Most rocks are solid, but a body of water is just as much a "rock" as is a body of ice. Most rocks are firm, but sand, gravel, and other loose masses are considered rocks by the geologist but not by the engineer. Most rocks are aggregates of two or more minerals, as is granite, but single minerals are rocks also when they occur on a large enough scale. Rock salt, which is composed solely of halite, is an example. Some examples bear the same name whether mineral or rock, such as anhydrite and gypsum. Most rocks consist of one or more crystalline minerals, but water is a liquid, coal is composed of organic matter not usually considered a mineral, and obsidian is natural, volcanic glass. All rocks with which we are familiar are terrestrial, but meteorites are rocks from outer space.

Aa. Rough, jagged lava, clinkery in appearance. This is a native Hawaiian word, now used everywhere for rough lava and the basalt that hardens from it. It is the opposite of *pahoehoe.*

Accessory mineral. A minor mineral in a rock. It occurs as less than 5 percent of the rock and is not involved in the naming of the rock, as is an *essential mineral.*

Acidic. Having a high content of silica. As the opposite of *basic,* this is a misleading term, but it is still widely used. Thus, granite, rhyolite, and obsidian are typical acidic rocks. Such rocks usually have a high content of alkalies and aluminum. Synonyms with special meanings include *acid, silicic, persilicic, felsic,* and *leucocratic.*

Adamellite. A visibly crystalline igneous rock composed of alkali feldspar and plagioclase feldspar (in about equal amounts) and quartz. It includes both quartz monzonite and granodiorite.

Agglomerate. A volcanic rock composed of large fragments imbedded in an ashy matrix.

Alaskite. A variety of granite in which the dark minerals are less than 5 percent.

Alteration. Physical or chemical changes in a mineral or rock after its original formation. Weathering is an important kind of alteration.

Amphibolite. A visibly crystalline igneous rock composed almost entirely of amphibole, usually hornblende.

Also, a metamorphic rock composed mostly of amphibole and plagioclase feldspar.

Amygdaloid. An igneous rock containing gas cavities. These may be filled or lined with secondary minerals. The rock is usually lava but it may occur also at the margins of other igneous bodies.

Andesite. A fine-grained igneous rock, the extrusive equivalent of diorite. The color is usually dark, often gray, brown, red, green, or nearly black. The individual minerals can seldom be identified at sight except when they occur as distinct grains (phenocrysts) in porphyritic varieties. Glass may be present in varying amounts. Andesite is abundant on both sides of the Pacific Ocean, notably in the Andes Mountains.

Anhedral. Lacking crystal faces. It applies to minerals in rocks and contrasts with *euhedral* and compares with *subhedral*.

Anhydrite. A sedimentary rock composed of the mineral anhydrite. It is deposited by the evaporation of sea water. It is usually white unless stained by chemical impurities.

Anorthosite. A variety of gabbro composed almost entirely of plagioclase feldspar. It may be light or dark depending on the feldspar present, which is usually labradorite. It occurs with valuable iron-ore deposits of magnetite in New York and is also found in large bodies in Norway, Labrador, Quebec, Minnesota, Montana, and South Africa.

Anthracite. A high rank of coal, occurring in sedimentary beds.

Aphanite. An igneous rock consisting of less than 50 percent visibly crystalline minerals. Aphanites include felsite, rhyolite, andesite, and numerous other extrusive rocks. Many examples are porphyritic.

Aphanitic. Fine grained, applied to rocks. The individual constituents cannot be identified by the eye alone. *Felsitic* means the same thing, as applied to light-colored rocks.

Aplite. A rather fine-grained igneous rock composed of quartz and alkali feldspar (orthoclase or microcline). The typical texture is almost sugary. Aplite is complementary in grain size to pegmatite and occurs with it as its opposite phase.

Apophysis. An ore vein or rock dike or sill that is an offshoot from a larger body.

Arenaceous. Sandy. A sandy shale is thus said to be arenaceous.

Arenite. A medium-grained sedimentary rock composed of fragmental material. It includes sandstone, arkose, and graywacke.

Argillaceous. Clayey. Thus, we speak of an argillaceous sandstone.

Argillite. A hard, fine-grained sedimentary rock. It resembles slate but is not capable of being split.

Arkose. A sedimentary rock containing considerable feldspar. Either sandstone or conglomerate can be an arkose, or arkosic.

Ash. Volcanic material of small size, under 4 millimeters. Still smaller particles (under 1/4 millimeter) constitute *volcanic dust.* When cemented, ash is called *tuff*.

Aureole. The zone of metamorphosed rock around an igneous intrusion.

Basalt. A fine-grained igneous rock, the extrusive equivalent of gabbro. The color is usually dark gray or black. The individual minerals can seldom be identified at sight, except when they occur as distinct grains *(phenocrysts)* in porphyritic varieties. Glass may be present in varying amount. Basalt is the principal rock underlying the oceans and makes up most of the world's vast lava flows, as in India, the Columbia Plateau (covering parts of five Western states), the Paraná Basin (South America), Iceland, and Hawaii. *Columnar basalt* shows a joint structure characteristic of this rock.

Basic. Having a low content of silica. As the opposite of *acidic,* this is a misleading term, but it is still widely used. Thus, gabbro and basalt are typical basic rocks. Such rocks usually have a high content of iron, magnesium, and calcium. Synonyms with special petrologic meanings include *subsilicic, mafic, ferromagnesian, femic,* and *melanocratic. Ultrabasic* (or *ultramafic)* rocks contain conspicuous amount of metallic minerals.

Bauxite. A sedimentary rock composed of various aluminum silicate minerals. It is almost the sole commercial source of aluminum.

Bauxite

Bedding. The layering of rocks. This applies to most sedimentary rocks and to lava. The boundary between the individual beds, layers, or strata is called the *bedding plane.* Its relative porosity may make the bedding plane a suitable place for the deposition of minerals. The study of layered rocks is called *stratigraphy.*

Bedrock. The solid rock underlying soil or loose rock.

Bentonite. A sedimentary rock consisting largely of the clay mineral montmorillonite. It is formed by the chemical alteration of volcanic ash deposited in water. It then has a high capacity for taking up water and swelling correspondingly. Its various industrial uses depend on its swelling properties and plasticity. Fresh bentonite has a white or a pale tint, changing to cream or even brown upon weathering.

Bituminous coal. A medium rank of coal, between lignite and anthracite.

Blanket deposit. A horizontal ore deposit that is larger and wider than it is thick.

Boss. A dome-shaped body of igneous rock exposed by erosion.

Bostonite. A fine-grained variety of syenite, composed of albite and microcline (feldspar) and a little pyroxene. It has a sugary texture.

Breccia. A fragmental rock composed of angular fragments.

Brecciation is the breaking of rock into such fragments. Breccias of several origins are known. Unless otherwise stated, a breccia is considered sedimentary, the opposite of rounded conglomerate. There is also *fault breccia* (due to friction along a fault), *talus breccia* (due to gravity movements along a cliff), *intrusive* and *extrusive volcanic breccia* (due to gas explosions in a pipe or volcano), and *flow breccia* (due to fragmentation within a lava flow).

Calcareous. Containing calcium carbonate. Thus, we speak of a *calcareous shale.*

Camptonite. Hornblende melaphyre. It is associated with bodies of syenite.

Cement. The bonding matter of sedimentary rocks. It is most often silica, calcium carbonate, iron oxide or hydroxide, gypsum, or barite. The color of many sedimentary rocks is due to the color of the cement rather than to that of the principal minerals.

Cement rock. Limestone containing clay. The mixture is in the right proportions to be used to make cement.

Chalk. A soft, fine-grained variety of limestone. It is usually composed of microscopic shells.

Cinders. Solid volcanic material having a range of size larger than ash or dust and smaller than blocks or bombs. It is the same as *lapilli.*

Clastic. Fragmental. Clastic sedimentary or igneous rocks are those consisting of individual particles cemented together.

Claystone. A sedimentary rock composed of clay-sized particles. It differs from shale in not showing a thin bedding; instead it tends to break into blocks.

Coal. A sedimentary rock composed of altered plant remains. It originates in peat and goes through the ranks of lignite, bituminous coal, and anthracite, becoming graphite if conditions are favorable.

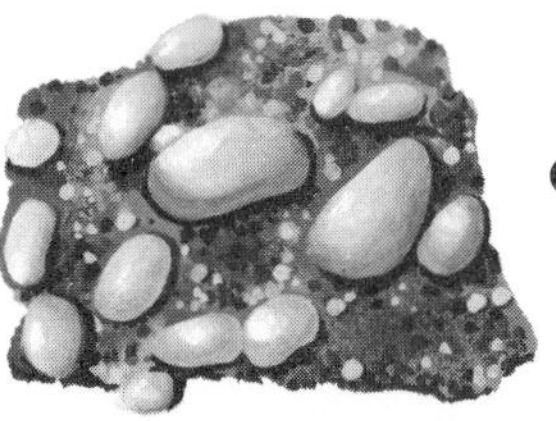

Conglomerate

Conglomerate. A sedimentary rock composed of rounded gravel. When the particles are angular, the rock is a *sedimentary* *breccia*. As the particles become smaller, conglomerate grades into sandstone. The material may be individual mineral grains or fragments of rock. The color varies with the constituents and the natural cement.

Coquina. A variety of limestone composed of organic fragments. Shells make up the principal material, but crinoid stems and plates, pieces of coral, and algae are also abundant. The coarse coquina of Florida is of widespread distribution.

Coral. A sedimentary rock composed of calcium carbonate secreted by corals and other organisms. It is a reef limestone built up as a solid body. Certain colorful varieties–red or pink–are used as gems.

Dacite. A fine-grained igneous rock, the extrusive equivalent of quartz diorite or tonalite. It is generally gray. The individual minerals can seldom be identified at sight, except when they occur as distinct grains (phenocrysts) in porphyritic varieties. Glass may be present in varying amounts.

Detrital. Consisting of rock or mineral fragments. Detrital sedimentary rocks are those consisting of individual particles cemented together.

Diabase. A fine-grained igneous rock composed of visible grains of labradorite (plagioclase) feldspar and augite (pyroxene). It is intermediate in texture between gabbro (coarse) and basalt (fine) but has the same mineral composition. In Great Britain this rock is called *dolerite*.

Diatomite. A sedimentary rock composed of diatom shells. Diatoms are one-celled plants, abundant in marine and fresh-water plankton, that absorb silica and secrete shells of the same material, which accumulate to form this rock. *Diatomaceous earth* is another name for this material, which is mined in California and is used in making insulators, abrasives, and filters.

Diorite. A visibly crystalline igneous rock composed of plagioclase feldspar in excess of alkali feldspar. The color is usually dark gray because of the dark minerals always present (mostly hornblende but also biotite or pyroxene).

Disseminated. Scattered through a rock. Diamond crystals, for example, are disseminated in kimberlite.

Dolerite. A medium-grained variety of gabbro. It is prominent in thick lava that cooled slowly. Another name is *microgabbro*.

Dolomite. A sedimentary rock composed of the mineral dolomite. It is deposited either by the evaporation of sea water, or by its replacement of limestone.

Dripstone. Cavestone deposits in a cavern. These include *stalactites, stalagmites, columns, pillars,* and *helictites* of calcite, aragonite, and other minerals.

Druse. A small cavity in rock, lined with small (or *drusy)* crystals.

Dunite. A variety of peridotite consisting almost entirely of olivine.

Eclogite. A visibly crystalline rock composed of garnet and pyroxene. The association of pink garnets (almandite, pyrope) and grass-green pyroxene (omphacite) makes an attractive combination. This rock is more often metamorphic than igneous.

Erratic. A glacial boulder lying on rocks of a different type. The small erratics of the north-central states include diamonds, probably from Canada, and pieces of native copper from the Keweenaw Peninsula of Michigan.

Essential mineral. A major mineral in a rock. It is taken into account in naming the rock, as hornblende in hornblende granite, and (unlike an accessory mineral) it usually occurs as more than 5 percent of the rock.

Euhedral. Having complete crystal faces. It applies to minerals in rocks and contrasts with *anhedral* and compares with *subhedral.*

Evaporite. A sedimentary rock deposited by the evaporation of mineral-bearing water. Examples include rock salt, limestone, anhydrite, and gypsum.

Exposure. An outcrop of igneous rock.

Fanglomerate. Breccia of sedimentary origin.

Felsite. A fine-grained igneous rock composed of light-colored minerals, with or without glass. This is a field term, useful because the individual grains can seldom be identified at sight, although they are usually quartz and alkali feldspar. Felsite includes rhyolite, trachyte, latite, quartz latite, dacite, phonolite, and (when light-colored) andesite.

Felsitic. Fine grained, applied to light-colored rocks. The individual constituents cannot be identified by the eye alone. *Aphanitic* means the same thing but is not restricted to light-colored rocks.

Felsophyre. Light-colored (or *felsite)* porphyry.

Flow banding. Alternate layering of different materials in an igneous rock. It is especially noticeable in lava flows of high silca content and is, to collectors, most familiar in volcanic glass as "double-flow" obsidian.

Fluvial deposit. Sediment deposited by a stream. It has been transported as particles from elsewhere and may with consolidation become a sedimentary rock.

Foliation. A parallel texture in metamorphic rock. It is caused by the segregation of different minerals into bands or layers as a result of directed pressure. Foliation is about the same as *schistosity*.

Fragmental. Consisting of individual particles cemented together. Fragmental, or *clastic,* rocks can be sedimentary or igneous.

Friable. Crumbly. Even a sandstone that has quartz (hardness 7) as its chief or sole mineral may, because it is weakly cemented, be easily pulled apart with the fingers. Shale is almost always friable.

Gabbro. A visibly crystalline igneous rock composed of plagioclase and dark minerals (especially pyroxene). The color is therefore dark. Gabbro occurs in medium-sized bodies. Anorthosite and norite are important varieties of gabbro.

Geyserite. A deposit of silica at hot springs and geysers. It is also called *siliceous sinter.*

Biotite gneiss from Massachusetts, a typical metamorphic rock. Bands of feldspar and quartz alternate with bands of dark minerals—the biotite mica being the predominantly dark mineral here.

Gneiss. A coarsely banded metamorphic rock. Typically, bands of feldspar and quartz alternate with bands of dark minerals. Most varieties of gneiss are named according to the dominant mineral, of which the more common include *biotite gneiss, hornblende gneiss,* and *garnet gneiss.* A combined name, such as *granite gneiss,* indicates the general composition of the original source of the rock, if known. The color varies correspondingly. *Augen gneiss* shows eyelike fragments or crystals of feldspar or quartz or both.

Grain. An individual material of which a rock is composed. It may be either a mineral or a piece of glass. A grain may also be a loose particle in a sediment, which may then include pieces of rock.

Granite. A visibly crystalline igneous rock composed mainly of quartz and alkali feldspar (orthoclase or microcline). It is the best known igneous rock and constitutes more than 90 percent of the rocks of the continents. The color, which is governed chiefly by the kind of feldspar present, is most often light gray, white, or mottled pink. The chief varieties of granite contain biotite, muscovite, amphibole, or pyroxene.

Granodiorite. A visibly crystalline igneous rock composed of alkali feldspar and plagioclase feldspar (in about equal amounts) and quartz. Flow structure is a common feature. This rock is associated with many rich mineral deposits in western North America. It is less abundant than granite and usually a little darker gray.

Granular. The texture of igneous rocks that consist of equal-sized mineral grains. Granite has a typically granular texture.

Graphic granite

Graphic granite. A variety of pegmatite composed of an intergrowth of microcline feldspar and rods of quartz. The quartz is grooved and is distributed in a distinctive angular pattern that resembles ancient writing.

Graywacke. A sedimentary rock composed of quartz, feldspar, and dark rock fragments.

Greenstone. A metamorphic rock consisting largely of green minerals. These are mostly chlorite, epidote, and hornblende. Canada has a large amount of greenstone.

Groundmass. The finer-grained background in a porphyry. In it are set the phenocrysts.

Gypsum. A sedimentary rock composed of the mineral gypsum. It is deposited by the evaporation of sea water. It is usually white unless stained by chemical impurities.

Hornblendite. An intrusive igneous rock composed almost entirely of hornblende. The color is dark green or black.

Hornfels. A metamorphic rock produced by heat. It is typically a sugary-grained rock occurring in contact-metamorphic zones. It frequently contains large crystals, as in the so-called spotted slate.

Igneous. Solidified from a molten state. If formed beneath the the surface, from magma, igneous rock is *intrusive.* If formed near or upon the surface, from lava, it is *extrusive.* Igneous rock is presumed to be the original of all rock, and it may later change to sedimentary or metamorphic rock. Common igneous minerals include quartz, feldspar group, feldspathoid group, mica group, amphibole family, pyroxene family, and olivine. Important accessory minerals in igneous rocks include zircon, sphene, apatite, rutile, corundum, garnet group, magnetite, ilmenite, hematite, and pyrite.

Ironstone. A sedimentary rock containing a substantial proportion of iron minerals. These are mainly siderite and chamosite; hematite and glauconite also occur, often with clay in a mixture usually called *clay ironstone.*

Itabirite. A metamorphic rock composed mostly of quartz and micaceous hematite. It has a glistening luster and resembles mica schist.

Jaspilite. A sedimentary rock consisting of alternating layers of red jasper and hematite. Associated with taconite, it is abundant in the Lake Superior region of iron-ore deposits.

Kimberlite. A weathered variety of peridotite. It is broken (brecciated) and chemically altered. This is the diamond-bearing rock of South Africa and Arkansas; it is found elsewhere without diamond.

Lamprophyre. A dark igneous rock occurring in dikes. When it contains light-colored minerals, these are present only in the groundmass. Lamprophyres are found most often with syenite and diorite.

Lapilli. Solid fragments of volcanic material having a range of size larger than *ash* or *dust* and smaller than *blocks* or *bombs.* It is the same as *cinders.*

Larvikite. A variety of syenite, composed of gemmy orthoclase (feldspar), it has a rich blue sheen resembling that of labradorite. The source of this valuable building stone, often sold as *Norwegian pearl-gray granite,* is Larvik, Norway.

Laterite. A residual deposit of hydrous aluminum and iron oxides. It develops typically in the tropics.

Latite. A fine-grained igneous rock, the extrusive equivalent of monzonite. The color is generally gray. The individual minerals can seldom be identified at sight, except when they occur as distinct grains (phenocrysts) in porphyritic varieties. Glass may be present in varying amounts.

Leucophyre. A light (or *felsite)* porphyry.

Lignite. A low rank of coal, between peat and bituminous coal.

Limestone. A sedimentary rock composed of calcite. It is formed by the accumulation of shells, limy sand, or mud, and the deposition of calcium carbonate. Chalk and coquina are special varieties of limestone, and there are others going under variant names, such as *lithographic limestone.*

Lithology. The study of rocks, especially sedimentary rocks.

Loess. A sedimentary rock deposited by wind. The material is yellow or brown sand and silt, often of glacial origin. It lacks bedding and tends to stand as cliffs.

Lutite. A fine-grained sedimentary rock composed of tiny fragments. It includes siltstone, shale, mudstone, claystone, argillite, and loess.

Marble. A metamorphic rock composed of calcite or dolomite. It is thus derived from the sedimentary rocks, limestone or dolomite. The original, white color of the pure rock may be supplanted by the presence of impurities or newly developed minerals, such as serpentine, graphite, or hematite. Fine marble from such countries as Greece, Italy, France, and Belgium has long been famous for statuary, monument, and building stone.

Marl. A sedimentary rock composed of a mixture of a carbonate material and clay.

Massive. A thick or unoriented rock. A massive rock either occurs in thick beds (as a sedimentary rock or lava flow) or else seems uniform in structure, without definite top or bottom or other directional features.

Megascopic. Visible to the unaided eye. *Macroscopic* means the same but is apt to be confused with *microscopic,* the opposite term.

Melaphyre. Dark porphyry.

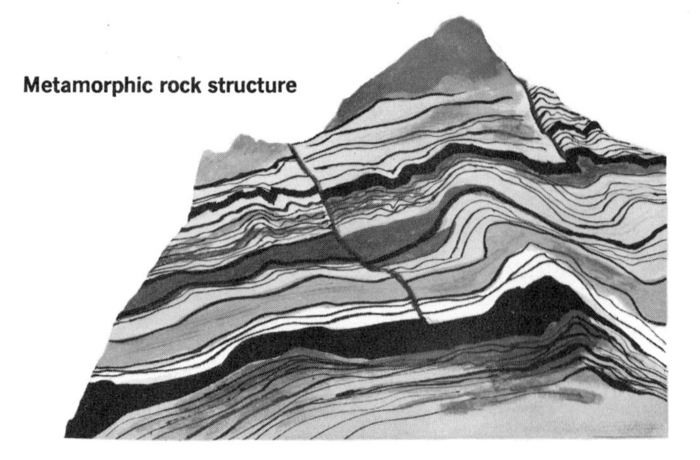
Metamorphic rock structure

Metamorphic. Changed from a previous igneous or sedimentary state by the action of heat, pressure, and fluid. Familiar metamorphic rocks include gneiss, schist, phyllite, slate, marble, quartzite, and serpentine. *Metamorphism* is a major geologic process. It may take place next to magma *(contact metamorphism)* or over a large area *(regional metamorphism),* occurring in zones based on the intensity of metamorphism and marked by different mineral associations. Its effect upon rocks is to create new minerals and produce new textures. The minerals are either recrystallized from others of the same chemical composition, as when the calcite of limestone changes to the calcite of marble; or are recombined to make new chemical compounds, as when the silica of quartz (SiO_2) and the calcium carbonate of calcite ($CaCO_3$) together make wollastonite ($CaSiO_3$), setting free carbon dioxide (CO_2). Common metamorphic minerals include quartz, feldspar group, mica group, amphibole family, pyroxene family, garnet group, calcite, sillimanite, kyanite, staurolite, serpentine, and chlorite.

Miarolitic cavity. A large cavity in pegmatite. It may also occur in other granitic rocks, but the ones found in pegmatite are famous for their fine crystals, which may project inward toward the center or entirely fill the cavity.

Microgabbro. Another name for dolerite.

Monzonite. A visibly crystalline igneous rock composed of alkali feldspar and plagioclase feldspar (in about equal amounts). Most monzonite is light gray to medium gray. Closely spaced joints and many inclusions are common features. Monzonite usually forms bodies of medium size. In the western United States it is associated with numerous ore deposits.

Mudstone. A sedimentary rock composed of mudlike particles. It differs from shale in not showing a thin bedding; instead it tends to break into blocks.

Mylonite. Metamorphic rock produced by crushing along a fault surface.

Nepheline syenite. A visibly crystalline igneous rock composed of alkali feldspar and nepheline. The color is usually white to gray, but dark minerals may occur in concentrations. This is an unusual rock containing many rare accessory minerals.

Norite. A variety of gabbro composed of labradorite (plagioclase) feldspar and hypersthene (pyroxene). It is a dark rock associated with the valuable ore deposits of platinum, nickel, and copper at Sudbury (Ontario).

Novaculite. A variety of metamorphosed chert. It comes from Arkansas and is used for whetstones.

Obsidian. Volcanic glass. Most obsidian has the same chemical composition as rhyolite or granite. The color is black or (less often) brown, red, or green, and frequently banded, as in "double-flow" obsidian. Carvings, mirrors, weapons, and tools have long been made from this rock, which has a worldwide distribution wherever volcanic activity has taken place. *Silver-sheen* and *golden-sheen* obsidian are popular lapidary varieties. Lumps known as *Apache tears* are familiar to collectors. With added water, obsidian grades into perlite and pitchstone. Slaggy-looking obsidian grades into rhyolite.

Snowflake obsidian—black, volcanic glass with inclusions of cristobolite. This is a favorite material for the lapidary as it makes handsome sets for silver jewelry. This specimen is from Millard County, Utah.

Oil shale. A variety of shale containing carbonaceous matter that can be distilled to make oil.

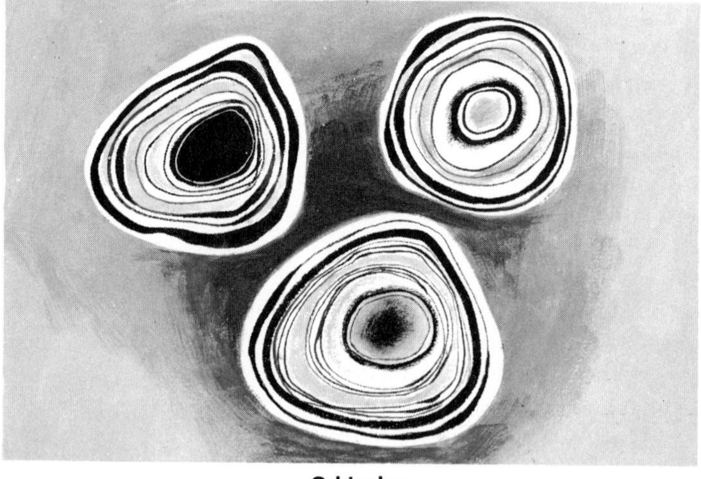

Orbicules

Orbicular. A rounded structure. This spheroidal shape occurs in certain rocks, such as granite. The aggregates of minerals are usually zoned in such concentric shells.

Organic rock. A rock produced by plant or animal life, such as coal and petroleum.

Outcrop. A surface exposure of sedimentary rock or mineral deposit.

Pahoehoe. Smooth, ropy lava of various patterns and textures. This native Hawaiian word is now applied everywhere to smooth lava and basalt that hardens from it. It is the opposite of *aa*.

Paragenesis. The association of minerals in a rock and the order in which they formed.

Peat. A sedimentary rock composed of plant matter that has begun to alter to coal.

Pegmatite. A very coarse-grained igneous rock. It forms in veins, lenses, and other shapes, and is most often associated with granite. The minerals are those of ordinary granite (quartz, feldspar, mica), to which may be added abundant tourmaline, beryl, or topaz. Certain pegmatite is noted for its rare and gem minerals, and bodies of this kind are the collector's favorite rock. Cavities in pegmatite may contain crystals of large size and high value, projecting toward the center or filling the opening completely. The world's largest crystals, as much as several thousand tons in weight, occur in pegmatite. However, a variable grain size is even more characteristic of pegmatite than a large size, as is typified in the variety called graphic granite.

Pele's hair. A threadlike form of volcanic glass. It is produced by the blowing-out of lava by explosion or by bursting bubbles on lava lakes, as in Hawaii. Most of it is basalt. It is named after Pele, Hawaiian goddess of the volcano.

Pele's tears. Drops of volcanic glass. They are thrown from volcanoes during eruptions and are common in Hawaii. Most of them are basalt.

Peridotite. A visibly crystalline igneous rock composed mostly of olivine. An altered and brecciated variety is kimberlite.

Perlite. A volcanic glass with rounded, pearly cracks. In the shells thus formed may be found pellets of obsidian. Perlite usually has the same chemical composition as rhyolite but with added water, though less than that in pitchstone.

Petrography. The description and classification of rocks. It is a part of *petrology*.

Petrology. The study of rocks. It includes *petrography* (the description and classification of rocks) and *petrogenesis* (the origin and history).

Phaneritic. Visibly crystalline. Igneous rocks having a phaneritic texture may range from a rather fine to an extremely coarse (pegmatitic) grain size, but the mineral grains are all crystalline (not glassy) and can be seen without a microscope.

Phenocryst. A larger grain in a porphyry. The phenocrysts are set in the groundmass and are believed to have formed first.

Phonolite. A fine-grained igneous rock, the extrusive equivalent of nepheline syenite. The color is usually light gray. The individual minerals can seldom be identified at sight, except when they occur as distinct grains (phenocrysts) in porphyritic varieties. Phonolite is the chief rock in the gold deposits at Cripple Creek (Colorado); it is also found along the Rhine River.

Phosphorite. A sedimentary rock composed of phosphatic material. This varies in form but, as *collophane,* it has the general composition of apatite. Phosphorite originates by both organic and inorganic processes. It may be almost pure but is generally mixed with enough other material to form phosphatic sandstone, shale, or limestone.

Phyllite. A finely banded metamorphic rock. It is intermediate between schist and slate. The color is typically dark, but muscovite mica or chlorite gives phyllite a silky sheen.

Pitchstone. A pitchy-looking volcanic glass. The water content is rather high (4 to 10 percent). The color, in a range of shades, may be uniform or mottled or streaked.

Pocket. A cavity in rock, containing crystals or other minerals. A small cavity, containing small *(drusy)* crystals, is called a *druse.* The large pockets in pegmatite, so well known as a source of fine crystals, are called *miarolitic cavities.*

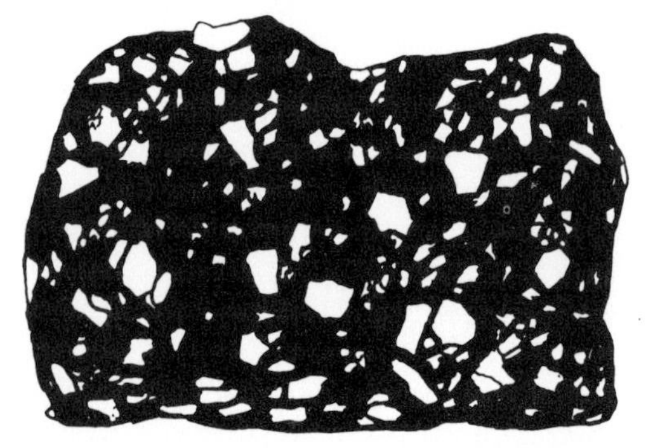

Porphyry

Porphyry. An igneous rock consisting of conspicuously different grain sizes. The larger grains are called *phenocrysts* and are embedded in the *groundmass,* in which they developed earlier or simultaneously by an unequal growth. In porphyritic texture, the phenocrysts may be any size but must be definitely coarser than the rest. Porphyry is a common miners' name for any ore-bearing rocks in western United States, but it is correctly used only as a descriptive term. It is given as *granite porphyry* (or *porpyhritic granite),* etc., according to the rock type and the proportion of phenocrysts to groundmass. A *porphyry* has more than half phenocrysts; a *porphyritic* rock has less than half.

Puddingstone. A popular name for conglomerate.

Pumice. A frothy rock erupted by a volcano. It is a natural glass filled with air bubbles, light enough to float on water. The color is usually white or light gray, and the chemical composition is like that of obsidian. Volcanic regions produce it everywhere, and it is used as an abrasive.

Pyroclastic. Broken by volcanic explosion. Pyroclastic rock has been ejected from a volcano. It may be pieces of country rock or earlier material of the volcano itself.

Pyroxenite. A visibly crystalline igneous rock composed almost entirely of pyroxene.

Quartz diorite. A visibly crystalline igneous rock composed of plagioclase feldspar, quartz, and hornblende or biotite or both. The color is generally gray. This rock forms side aspects of larger bodies. Another name for quartz diorite is *tonalite.*

Quartzite. A metamorphic rock composed mostly of quartz. Originally sandstone, it has been changed so completely that the cementing material is as resistant as the original mineral grains. The color is typically white. Quartzite is an extremely hard and durable rock.

Quartz latite. A fine-grained igneous rock, the extrusive equivalent of adamellite (or quartz monzonite or granodiorite). It is usually gray and is intermediate in composition between rhyolite and dacite. The individual minerals can seldom be identified at sight except when they occur as distinct grains (phenocrysts) in porphyritic varieties. Glass may be present in a varying amount.

Quartz monzonite. A visibly crystalline igneous rock composed of alkali feldspar and plagioclase feldspar (in about equal amounts) and quartz. Flow structure is a common feature of this rock, which is associated with many rich mineral deposits in western North America. It is less common than granite and is usually a little darker gray.

Rhyolite. A fine-grained igneous rock, the extrusive equivalent of granite. The colors are many, but most of them are light. The individual minerals can seldom be identified at sight, except when such as sanidine feldspar, quartz, plagioclase feldspar, biotite, or amphibole occur as distinct grains (phenocrysts) in porphyritic varieties. Glass is common in much rhyolite, which then grades into perlite, pitchstone, and obsidian. Topaz is an interesting mineral in the rhyolite of the Thomas Range, Utah.

Rock salt. A sedimentary rock composed of halite. It is deposited by the evaporation of sea water or concentrated lakes, and constitutes great salt domes, layers of various thicknesses, and isolated playas in arid regions. Beds of rock salt are often associated with strata of gypsum, anhydrite, and shale.

Rudite. A coarse-grained sedimentary rock composed of fragmental material. It includes conglomerate and breccia.

Sandstone (from Potsdam, New York) showing the gritty texture. Sandstone is one of the three most common sedimentary rocks in the earth's crust.

Sandstone. A sedimentary rock composed of sand-sized particles. The sand is usually quartz, but it may include a wide range of other substances, including heavy minerals, gypsum, or coral. The color varies with the constituents and the natural cement. As the grain size increases, sandstone grades into conglomerate. Sandstone is a common building stone.

Schist. A medium-banded metamorphic rock. With increasing coarseness it grades into gneiss, but a distinctive feature of schist is the general parallelism (lamellar arrangement) of the minerals. Most varieties of schist are named according to the dominant mineral, of which the more common include *mica schist, garnet schist, chlorite schist,* and *hornblende schist.* The color varies correspondingly.

Schistosity. A parallel texture in metamorphic rock. It is caused by the segregation of platy and oval mineral grains into layers as a result of directed pressure. Especially typical of schist, this structure is about the same as *foliation.*

Scoria. Dark, porous volcanic rock. Because of its basic composition, scoria is heavier than pumice.

Sedimentary. Formed either by the accumulation of fragments of older rock or by chemical precipitation from solution. The latter may be organic or inorganic. Familiar sedimentary rocks include conglomerate, sandstone, shale, limestone, dolomite, diatomite, gypsum, anhydrite, and rock salt. Sedimentary minerals are thus secondary, being derived from previous igneous, sedimentary, or metamorphic rocks. Common sedimentary minerals include quartz, feldspar group, halite, calcite, dolomite, gypsum, anhydrite, clay minerals, and hematite.

Segregation. Mineral matter accumulated in conspicuous aggregates.

Serpentine. A metamorphic rock composed mostly of the mineral serpentine. Veins of fibrous serpentine are *asbestos.*

Shale. A sedimentary rock composed of mud. This material has the grain-size of clay or silt and may be almost any color, though most often gray. Shale always shows a thin bedding; without it, the rock is called *mudstone* or *claystone.* It grades into sandstone by the addition of quartz, and into limestone by the addition of calcite. Shale is the most common sedimentary rock and also the weakest. It is used in making cement and plaster.

Siliceous sinter. Another name for geyserite.

Siltstone. A sedimentary rock composed of silt-sized particles. These range in size between sand and clay.

Slate. A fine-grained metamorphic rock having a sharp cleavage. The gray color is typical of slate, but it is not the only color. The sheets or slabs into which this rock splits are used for blackboards, roofing, and insulation.

Slaty cleavage. A parallel texture in metamorphic rock. It is caused by the segregation of platy and oval mineral grains into layers as a result of directed pressure. Especially typical of slate, this structure is about the same as that meant by *foliation* or *schistosity.*

Slickenside. A polished and grooved surface on rock, due to rubbing during movement along a fault.

Soapstone. A metamorphic rock composed mostly of talc. When impure, it is about the same as steatite.

Steatite. Impure soapstone.

Stockwork. A network of small ore veins that are mined as a unit.

Stratigraphy. The study of layered (bedded, stratified) rocks.

Stringer. A narrow vein of mineral or ore.

Structure. A large-scale feature of rock masses. These features include *bedding, jointing, cleavage, foliation, schistosity, brecciation,* and *flow banding.* They may also include *folding* and *faulting,* which are the major aspects considered in structural geology. The usual American distinction between structure and texture is often reversed in other countries, especially Germany. The terms are not firmly fixed.

Subhedral. Having some crystal faces. It applies to minerals in rocks and contrasts with *anhedral* and compares with *euhedral.*

Syenite. A visibly crystalline igneous rock composed of alkali feldspar. Various light colors are typical. Syenite is similar to granite but it has less than 5 percent quartz. Other minerals present generally include hornblende, biotite, and plagioclase feldspar. Its occurrence is less common than granite. Bostonite is a variety of syenite, as is nepheline syenite.

Tachylite. Basalt glass. Most volcanic glass, however, is obsidian.

Taconite. An iron-bearing, flinty rock capable of being used as an ore of iron. This low-grade, silica-rich rock is concentrated enough in the Lake Superior region, Wyoming, and elsewhere to yield a product suitable for the steel mills. Originally, taconite was the source of the now largely depleted high-grade hematite in Minnesota and adjoining states.

Tactite. A metamorphic rock formed along the contact of an intrusive igneous body and a carbonate rock. Garnet is the most typical mineral found in it.

Tephrite. A variety of basalt containing nepheline.

Texture. The small-scale features of rocks. These features include the size, shape, and pattern of arrangement of the constituents (rock particles, minerals, and glass) of which they are composed. The usual American distinction between texture and structure is often reversed in other countries, especially Germany. The terms are not firmly fixed.

Till. Glacial deposits unsorted by size or weight. When till is cemented naturally into solid rock it becomes *tillite.*

Tillite. A sedimentary rock composed of unsorted glacial deposits. The original, loose material is called *till.*

Tinguaite. A variety of phonolite occurring in dikes.

Tonalite. Another name for quartz diorite.

Trachyte porphyry, one of the more interesting-looking igneous rocks with its lath-shaped phenocrysts of potash feldspar.

Trachyte. A fine-grained igneous rock, the extrusive equivalent of syenite. Its color ranges from white to pink, green, or tan. The individual minerals can seldom be identified at sight, except when they occur as distinct grains (phenocrysts) in porphyritic varieties. Glass may be present in varying amounts.

Trap. A popular name for basalt and similar black or dark-green, fine-grained rock of lava origin.

Also, a closed rock reservoir containing oil or natural gas.

Travertine. A variety of limestone deposited by hot or cold springs. It is rather porous.

Tripoli. A sedimentary rock composed of silica in a rough, porous aggregate. It consists of fine shells or chert and is used as a filter and as an abrasive.

Tuff. A layered rock composed of volcanic ash or volcanic dust. Much of it has been deposited in bodies of water.

Ultrabasic. Having a very low content of silica. Ultrabasic rocks have a high content of iron and magnesium and reveal a conspicuous amount of metallic minerals. A synonym is *ultramafic*.

Unakite. A metamorphosed igneous rock composed of quartz, pink or red orthoclase feldspar, and abundant epidote. Unakite makes an attractive ornamental stone. The Blue Ridge Mountains, from Virginia to Tennessee, and the Great Smoky Mountains, between North Carolina and Tennessee, are the source of this rock.

Vesicle. A cavity in fine-grained or glassy igneous rock. It is caused by the expansion of escaping gas as the rock solidifies. When later minerals line or fill a vesicle, it becomes an *amygdale* or *amygdule*.

Vitrophyre. Porphyritic volcanic glass. The dark glass is usually pitchy looking, and the visible grains are most likely sanidine feldspar or quartz.

METEORITES

A meteorite is a rock from outer space. Meteorites are probably the remnants of a planet similar to the earth which was disrupted by internal explosion or external impact; they may, on the other hand, represent fragments of celestial bodies originally the sizes of asteroids. Upon entry into the atmosphere of the earth, they are heated to incandescence by air friction, producing a *meteor,* or "shooting star."

This meteorite of the siderite (nickel-iron) variety fell near Odessa, Texas. The "thumb marks" are quite evident in this specimen. Typical of a siderite, they are caused by surface-melting from air friction as the incandescent meteorite whirled through the earth's atmosphere.

Aerolite. A stony meteorite. Practically all aerolites show at least some trace of metal. Their chief constituents are silicate minerals, and so they resemble the basic rocks of the earth's crust.

Australite. A tektite from Australia.

Bediasite. A tektite from Texas, where it occurs near Bedias.

Billitonite. A tektite from Indonesia, which includes the island of Billiton.

Bottlestone. A popular name for moldavite.

Chondrule. A rounded grain embedded in a meteorite. Chondrules occur in stony (aerolite) meteorites, usually in a fragmented matrix. They are most often olivine or enstatite (pyroxene) but may be bronzite, hypersthene, augite (all pyroxene), plagioclase feldspar, carbonaceous or graphite, or glass.

False chrysolite. A popular name for moldavite.

Fusion crust. The outer layer of a meteorite, produced by friction during its passage through the earth's atmosphere. The fusion crust of a stony (aerolite) meteorite is glass, underlain by a mixture of glass and minerals. The fusion crust of a metallic (siderite) meteorite consists of magnetite.

Indochinite. A tektite from the area of former French Indo-China.

Mesosiderite. A variety of stony-iron (siderolite) meteorite. The metal is probably discontinuous. The other minerals are olivine, bronzite (pyroxene), and anorthite (plagioclase feldspar).

Micrometeorite. An extremely tiny meteoritic particle, too small to fuse during its passage through the earth's atmosphere.

Moldavite. A tektite from Rumania, which includes the former principality of Moldavia.

Multiple fall. A fall of two or more individuals broken from a single meteorite during its passage through the earth's atmosphere. It constitutes fewer individuals than a meteorite shower.

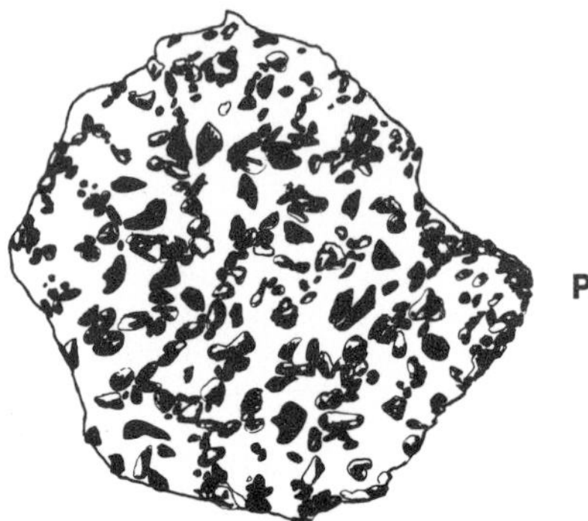

Pallasite

Pallasite. A prominent, beautiful variety of stony-iron (siderolite) meteorite. Large, rounded grains of olivine occur in a continuous, spongelike network of metal.

Rizalite. A tektite from the Philippines, where it occurs in the vicinity of Rizal.

Shower. A fall of numerous individuals broken from a single meteorite during its passage through the earth's atmosphere. Also, one of the periodic meteor displays prominent during certain times of the year, particularly in August and December.

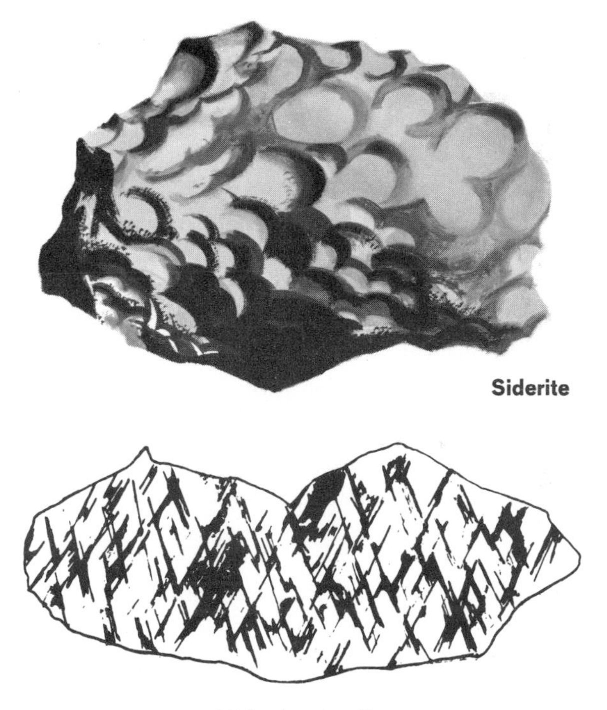

Siderite

Etched meteorite

Siderite. A metallic meteorite. Its most distinctive features are grooves, which may radiate from a single point, and pits known as "thumb marks." When etched with acid, the iron-nickel alloy usually shows fine lines (Neumann lines) or triangular crystal bands (Widmanstaetten figures).

Siderolite. A stony-iron meteorite. Siderolites includes pallasites and mesosiderites.

Silica glass. Glass of silica (SiO_2) composition, associated with meteorite craters and due to meteorite impact. Hence, it is also called *impactite.* It includes the varieties known as Darwin glass (from Tasmania, Australia), and Libyan glass (from the Libyan Desert).

Tektite. Natural glass of meteoritic origin. Named according to the place of discovery *(as moldavites, australites, billitonites).* Tektites are much alike internally. How and where they originated is not yet known. Perhaps the most interesting theory is that they are solidified droplets of lunar material, splashed into space by the impact of huge meteorites with the moon.

Water chrysolite. A popular name for moldavite.

Classification of Meteorites

Meteorites are generally put into three main classes, according to their relative proportions of metal and rock: *metallic* (nickel-iron alloys), *stony* (mostly silicate minerals), and *stony-iron* (both silicates and nickel-iron).

The classification most used today (by Prior after Rose-Tschermak-Brezina) is given in shortened form below.

I. Meteoritic irons—*siderites.* Consisting mainly of nickeliferous iron.

II. Meteoritic stony-irons—*siderolites.* Consisting of both iron and stony matter in large amounts.

III. Meteoritic stones—*aerolites.* Consisting mainly of stony matter—with nickeliferous iron and troilite, when present, scattered through it as small grains.

- (a) Chondritic meteoritic stones—*chondrites.* Containing chondrules.
- (b) Nonchondritic meteoritic stones—*achondrites.* Chondrules are absent.

The classification of Palache-Berman-Frondel is given below.

I. Iron meteorites—*siderites.* Consisting almost entirely of iron-nickel alloy.

II. Iron-stone meteorites. Consisting of iron and silicate minerals.

- (a) *Lithosiderites.* Consisting of a continuous, cellular matrix of iron-nickel alloy enclosing discontinuous bodies of silicate minerals.

(b) *Siderolites.* Consisting of a stony mass of silicate minerals within a discontinuous meshwork of iron-nickel alloy.

III. Stony meteorites—*aerolites.* Consisting of silicate minerals with clots, scales, or grains of iron-nickel alloy.

IV. Glassy meteorites—*tektites.* Consisting almost entirely of siliceous glass.

Meteorite Minerals

The minerals described below are not found in rocks of the earth but only in meteorites, except as indicated. The most abundant minerals in meteorites, apart from the nickel-iron alloy called kamacite, are enstatite, bronzite, and hypersthene (members of the enstatite series of the pyroxene group), and olivine. Also important are magnetite, chromite, other pyroxenes (pigeonite, hedenbergite, diopside, augite, clinoenstatite, clinobronzite, clinohypersthene), and plagioclase feldspar. Occasional minerals in meteorites include diamond, graphite, native sulfur, native copper, native gold, alabandite, pentlandite, chalcopyrrhotite, valleriite, chalcopyrite, pyrite, sphalerite, magnesite, calcite, dolomite, ilmenite, spinel, quartz, tridymite, cristobalite, apatite, gypsum, epsomite, bloedite, bravoite, zircon, and serpentine. Natural glass, various hydrocarbons, and amorphous carbonaceous matter are also present in various forms. New minerals, as yet unnamed, have been found in meteorites since 1960.

Cohenite—$(Fe,Ni)_3C$. A metallic-white, magnetic mineral found as grains or orthorhombic crystals only in meteorites, although some has been reported from native iron from Greenland.

Daubreelite—$FeCr_2S_4$. A black mineral found only in meteorites in small grains, plates, and veins.

Farringtonite—$Mg_3(PO_4)_2$. A colorless to wax-white to yellow mineral found only in meteorites.

Kamacite—Fe,Ni. A natural alloy of iron and 5.0 to 6.8 percent nickel, together with some cobalt and copper. It is metallic gray and magnetic. When intergrown with taenite, it constitutes *plessite.* These are the chief components of the siderite meteorites.

Lawrencite—$FeCl_2$. An unstable, green-brown mineral found only in meteorites. It absorbs water and becomes a greenish liquid which deposits a rusty stain.

Maskelynite–$(Na,Ca)(Al,Si)_4O_8$. An amorphous mineral resembling fused plagioclase feldspar, but found only in meteorites. It is clear, colorless, and vitreous.

Merrihueite–$(K,Na)_2(Fe,Mg)_5Si_{12}O_{30}$. A greenish-blue mineral found in 1965 as tiny aggregates in the Mezö-Madras meteorite.

Moissanite–SiC. A green mineral (of doubtful certainty) found as hexagonal crystals, but occurring only in meteorites.

Oldhamite–CaS. A water-soluble, brown mineral found as round grains, but occurring only in meteorites.

Osbornite–TiN. A golden-yellow mineral found as isometric crystals (octahedrons), occurring only in meteorites.

Schreibersite–$(Fe,Ni)_3P$. A metallic-white to yellow, magnetic mineral found as thin plates, shells, and needles, but only in meteorites. The mineral is also known as *rhabdite* when occurring in rods or needles. Specimens have also been reported as a product of combustion in coal mines in France.

Sinoite–Si_2N_2O. A light-gray mineral found in 1964 as irregular grains in the Jajh deh Kot Lalu meteorite.

Taenite–Fe,Ni. A natural alloy of iron and 13 to 48 percent nickel, together with some cobalt and copper. It is metallic-white and generally magnetic. When intergrown with kamacite, it constitutes *plessite*. These are the chief components of the siderite meteorites.

Troilite–FeS. A yellow to brown mineral found commonly in meteorites as rounded grains, aggregates, thin plates, and hexagonal crystals. The composition is related to pyrrhotite, a terrestrial mineral.

Ureyite–$NaCrSi_2O_6$. An emerald-green mineral found in 1965 as tiny grains in three meteorites. It is related to jadeite.

Interesting Meteorites

The following are meteorite falls and specimens noteworthy for their historical associations or great size.

Bacubirito, Sinaloa, Mexico. This, the third largest of the world's meteorites, weighs about 27 tons and measures 13 feet 1 inch, by 6 feet 2 inches, by 5 feet 4 inches. This siderite still lies where it was found in 1863.

Benares, Uttar Pradesh, India. This is the oldest house-striking meteorite on record, having fallen through the roof of a hut on December 19, 1798. Its weight is 2 pounds.

Benld, Macoupin County, Illinois. This aerolite penetrated the roof of a garage and buried itself in the seat cushion of an automobile on September 29, 1938. It weights 5 pounds and is in the Chicago Natural History Museum.

Camp Verde, Yavapai County, Arizona. A meteorite found about 1915, but not identified until 1939; it was wrapped in a feather blanket in an American Indian burial cyst. It weighed 135 pounds.

Canyon Diablo, Coconino County, Arizona. This prehistoric meteorite fall (20 to 75 thousand years old) produced, in the plateau country of northern Arizona, the best known of the world's meteorite craters. It has also yielded thousands of specimens since 1891.

Cape York, Greenland. The largest meteorite on exhibit, and the second largest in the world, is Ahnighito, known to the Eskimos by 1818 as "The Tent." Weighing 34 tons, this siderite measures 11½ by 7½ by 6 feet. The arctic explorer Robert E. Peary brought it to the American Museum of Natural History (New York City) in 1897, together with two smaller meteorites called "The Woman" (3 tons) and "The Dog" (½ ton).

Chupaderos, Jiménez, Chihuahua, Mexico. The two huge pieces of this siderite, 15 and 6½ tons, fell a few hundred feet apart—however, they fitted together perfectly, so they must have broken apart just before hitting the ground. The largest part ranks sixth in the world. Known for centuries, Chupaderos was first mentioned in 1852 and was taken to Mexico City in 1891.

Ensisheim, Alsace, France. This oldest dated fall—November 16, 1492—is represented by a 12-pound aerolite still preserved in the town hall.

Estherville, Emmet County, Iowa. A shower of meteorites followed an extremely brilliant meteor of May 10, 1879. The largest weighed about 437 pounds.

Forest City, Winnebago County, Iowa. An exceptionally brilliant meteor on May 2, 1890, was followed by a shower of aerolites, the largest recovered weighing 80 pounds.

Haviland, Kiowa County, Kansas. A crater-forming meteorite that has yielded many specimens since 1885.

Henbury, Northern Territory, Australia. A swarm of meteorites produced 13 craters within half a square mile. The discovery was made in 1931.

Hoba, Grootfontein, South-West Africa. The largest known meteorite, a siderite weighing about 60 tons, lies where it fell. It was found in 1920 and measures 9 by 9 by 3¼ feet.

Holbrook, Navajo County, Arizona. The smallest complete meteorite ever found, weighing 0.0183 grams, was found in 1940. It had fallen during the noted Holbrook meteorite shower of July 19, 1912, which yielded about 16,000 individual aerolites.

Johnstown, Weld County, Colorado. A shower of 35 aerolites disturbed a funeral procession on July 6, 1924.

Krasnsjarsk, Siberia. The study—by the German-Russian explorer, Peter Simon Pallas (1741–1811)—of this huge pallasite, found in 1749, was one of the milestones in the efforts of scientists to treat meteorites as natural objects from beyond the earth.

L'Aigle, Orne, France. This meteorite shower of about 2,500 individuals on April 26, 1803, finally settled the long dispute as to the authenticity of stones from the sky. The town is now usually spelled Laigle.

Long Island, Phillips County, Kansas. The second largest aerolite ever found, this yielded about 3,000 pieces that belonged to one stone that weighed about 1,244 pounds. They were found in 1891 in an area covering only about 100 square feet.

Mbosi, Tanganyika. This, the fourth largest of all meteorites, is estimated to weigh 25 to 27 tons. Its measurements are given as 13½ by 4 by 4 feet. This siderite lies where it was found in 1930.

Morito, Chihuahua, Mexico. This 12-ton siderite was known to the natives before 1600. It is the seventh largest in the world.

Mühlau, Innsbruck, Tirol, Austria. A 5-gram specimen is all that exists of the smallest total fall that has been recorded. This aerolite was found about 1877.

Norton County, Kansas. The largest aerolite ever found, weighing 1 ton, it is also the largest meteorite that has been recovered from a witnessed fall. The date was February 18, 1948.

Odessa, Ector County, Texas. Three craters were made by this meteorite fall, which was found before 1922.

Paragould, Greene County, Arkansas. The second largest areolite ever seen to fall, the date being February 17, 1930. After some pieces were removed, it was brought to the Chicago Natural History Museum, where it now weighs 745 pounds.

Pultusk, Warsaw, Poland. Perhaps 100,000 individuals made this the most spectacular meteorite shower on record. The date was January 30, 1868.

Sikhote-Alin, Siberia. On February 12, 1947, an outstanding fall produced 106 holes in the ground and yielded many thousand meteorites, the largest weighing nearly 2 tons.

Sylacauga, Talladega County, Alabama. This is the meteorite that broke through the roof of a house and injured a woman on November 30, 1954.

Tunguska River, Siberia. A spectacular impact on June 30, 1908, uprooted a forest and did other damage, but no meteorites have been recovered.

Willamette, Clackamas County, Oregon. The largest meteorite found in the United States, and the fifth largest in the world, weighs over 15½ tons and measures 16 feet 3½ inches, by 6 feet 6 inches, by 4 feet 3 inches. This siderite, found in 1902 and donated to the American Museum of Natural History (New York City) by Mrs. William E. Dodge, has deep cavities caused by rapid weathering.

The world's third largest meteorite, unnamed, was reported from the Gobi Desert in 1965. It is said to weigh 30 tons and to be on display in Urumchi, Sinkiang, China.

GEOLOGY AND MINING

Geology is the study and history of the earth. It treats of the materials that constitute the earth, the structures of which it is built, the surface features upon it, and the processes that operate to create and change them. It also covers the history of the earth, including (as fossils) the animals and plants that lived in the past, as recorded in the rocks. Its main divisions are *physical geology* and *historical geology.*

Mining consists of obtaining useful minerals from the earth. It includes both surface workings, as in a pit or quarry, and underground excavations.

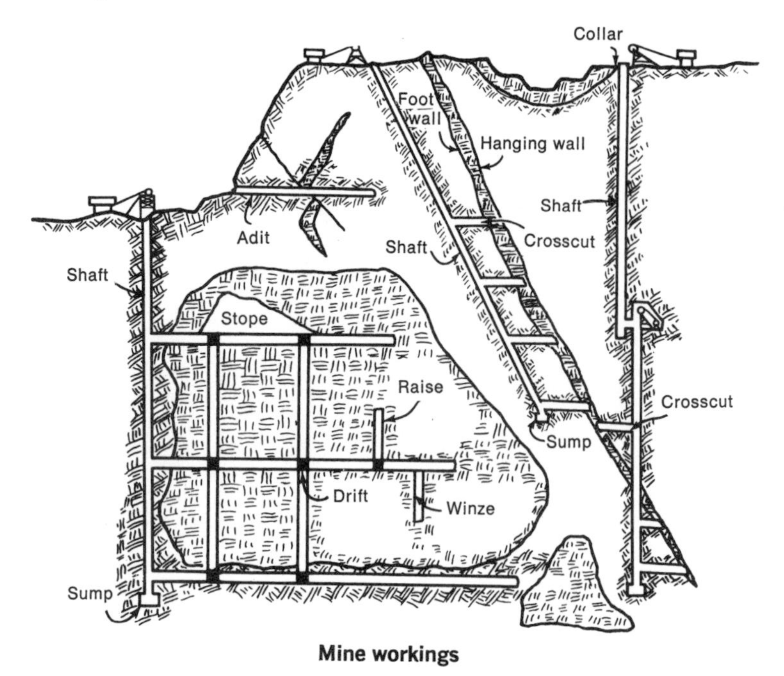

Mine workings

Barite. These shiny golden crystals from Gilman, Colorado, form a pleasing aggregation. They are associated in this locality with zinc ore.

Barite on calcite. Elongated, sea-blue crystals of barite show well the quality of the material that comes from Sterling, Colorado, one of the best-known localities for this color.

Crocoite (a chromate). These translucent, bright-red crystals, slender and separate, are from Tasmania, the most noted locality for fine crocoite.

Linarite. The intensely rich color of these little crystals represents one of the unusual hues of blue in the world of minerals. This vivid specimen is from Mexico.

Selenite rose. This flowerlike arrangement of crystals from Canon City, Colorado, is one of the most popular varieties of gypsum.

Celestite with sphalerite. A study in contrasts, with its cascade of bluish-white celestite set against the smouldering red fire of the sphalerite. This specimen comes from Ohio.

Celestite on calcite. Woodville, Ohio, has furnished collectors with this interesting association of white celestite on calcite.

Brochantite.

A highly magnified
view of this specimen
from Bingham, New Mexico,
shows us how its fine,
radiating crystals form into
little, bright, fuzzy balls.

Selenite.

This bristling cluster
of slender crystals with
their darkened tips
is from northern Michigan.
Fine lighting illuminates
them to best advantage
and shows their brilliant clarit

MOLYBDATES

Wulfenite twins. Contrast these comparatively thick crystals from Mexico with the more usual thinner habit. This close-up view shows the resinous luster characteristic of wulfenite.

Wulfenite. The very thin, tra parent, yellow-green variety this fragile cluster displays a other color variation and dime sion of this much desired spe men mineral. From Mexico.

Wulfenite, cerussite, and murdochite.

A striking trio of associated minerals from Tiger, Arizona. Bright-yellow to orange-red wulfenite, starlike clusters of white cerussite, and sparkling, dark murdochite.

Legrandite (an arsenate).

Smooth, yellow prisms with lustrous, flattened termination faces provide a pleasing contrast to their rusty, crusty background. From Mexico.

Wulfenite. These crystals, from Tiger, Arizona, show wulfenite at its best. The beveled edges are characteristic. The square wafers, attached to their matrix by one corner only, appear to have been stuck in place rather than having grown that way.

Pyramidal wulfenite. A rare crystal form of wulfenite is shown here in this exquisite golden crown from the Glove Mine in the Santa Rita Mountains of Arizona.

Wulfenite and pink calcite. The most typical wulfenite crystals, such as those below, often look like wafers of butterscotch candy. Note also the delicate, pink calcite crystals. From Mexico.

Aquamarine in quartz. A highly prized collectors' item is this specimen of fine gem beryl from the pegmatite deposits at Topsham, Maine. Silicates, the most numerous and complex of all mineral classes, include some of the most highly prized gems.

Dioptase on cerussite.

Golden cerussite and rich, velvety-green dioptase. Tiger, Arizona, is the home of this specimen.

Dioptase, cerussite, wulfenite, and chrysocolla.

This group of four minerals presents a study in unlike crystal habits. The dioptase is dark green, the cerussite is colorless, the wulfenite is golden, and the chrysocolla is blue and nearly white. From the Tiger Mine, Arizona.

Amazonstone with smoky quartz.

A frequent and favorite mineral combination from the Pikes Peak region; note the two white-tipped feldspar crystals.

Actinolite. This specimen is called *asbestiform actinolite* or *radiated actinolite*. When the fibers are more loosely held together, the mineral is called *asbestos*. From the Calumet mine, Salida, Colorado.

Epidote. From Salida, Colorado, comes this prized crystal. Note the yellowish- or pistachio-green hue; *pistacite* is an alternate name.

BLACK-LIGHT FLUORESCENCE. Green: Willemite. Red: Calcite. Yellow: Wernerite. Blue: Fluorite. Chartreuse: Calcium larsenate. Many minerals fluoresce—but not in their daylight colors. The fluorescence generally varies from specimen to specimen.

Abrasion. The process of rubbing or wearing away by friction. Thus, abrasion of rock is done by glaciers or by sediment in streams.

Abyssal. Formed by solidification of magma deep within the earth. Abyssal, or *plutonic,* bodies include those, such as *batholiths* and *stocks,* that have no visible floors. Abyssal rock is completely crystalline and typically coarse grained.

Accretion. The process by which minerals and rocks grow larger. Fresh material is added to the outside by accumulation of fragments or by precipitation of chemicals from solution.

Adit. A horizontal passage in a mine, open at the surface at one end only. If open at both ends, it is a *tunnel.*

Aeolian deposit. Sediment deposited by wind. It has been transported as particles from elsewhere and, by consolidation, may become a sedimentary rock. Sand dunes and loess are examples of aeolian deposits.

Aggradation. Depositional processes. These build up the surface of the land by the accumulation of sediment deposited by water, ice, and wind. The opposite is *degradation.*

Alluvial deposit. Sediment deposited by a stream. It has been transported as particles from elsewhere and may, by consolidation, become a sedimentary rock.

Alluvium. Sedimentary material carried and deposited by a stream. It may accumulate in a river bed, flood plain, or delta.

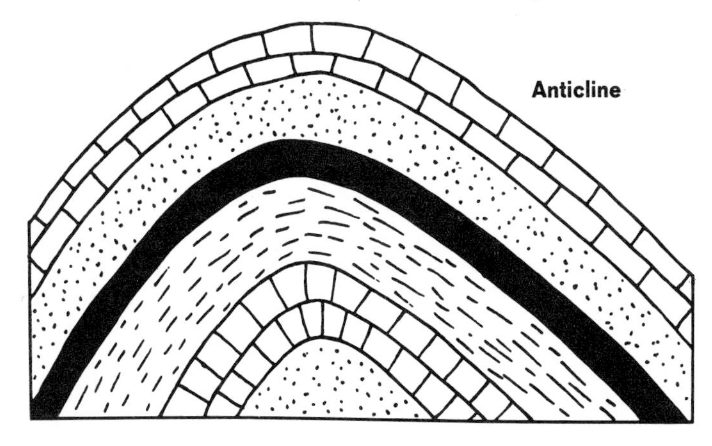

Anticline. A fold in which the rocks have been arched upward. Petroleum and natural gas often occur in the crests of anticlines.

Apex. The beginning, at the surface, of a vein.

Archeozoic. The oldest era of geologic time, marking the first part of Pre-Cambrian history. Rocks of this age are virtually free of fossils. They can be identified by their field occurrence and by radioactive dating.

Attrition. The process of rubbing or wearing away by friction. Thus, stream sediment causes attrition of solid rock.

Banded. Having thin, parallel zones of different colors or composition.

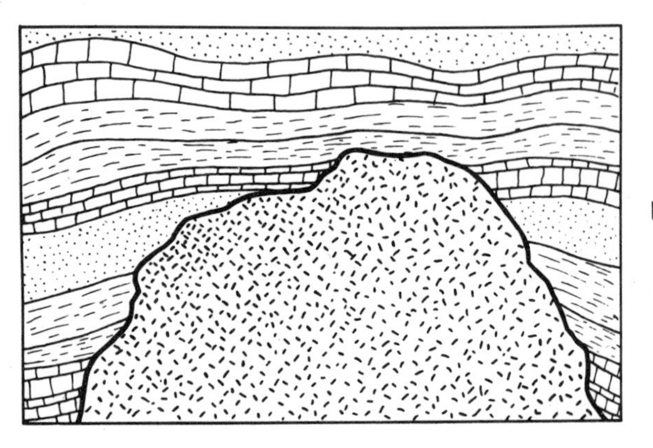

Batholith

Batholith. The largest bodies of intrusive igneous rock; exposed batholiths cover an area of more than 40 square miles. A *stock,* which may actually be the upper part of a batholith, is less in area than 40 square miles. Some batholiths, such as the Coast Range batholith of British Columbia, exceed 1,000 miles in length. The original source of most minerals is in batholiths, from which mineral-bearing solutions may be expelled and from which solid deposits may be eroded.

Bedded deposit. A mineral deposit in layered rock. It typically lies horizontal or nearly so.

Bonanza. A body of rich ore.

Buried placer. A former placer deposit that has become covered by a lava flow or other rock. This type of gold deposit is common in California.

Calcification. Petrification by means of calcium carbonate, $CaCO_3$. Calcite or aragonite is added to—or replaces—the original organic matter of the fossil. Thus, shell and bone are often changed to calcite.

Cambrian. The oldest period of the Paleozoic Era. It was named from the Latin name for Wales, where rocks of this geologic age were first studied. They contain fossils of every major invertebrate type.

Carbonation. Petrifaction (petrification) by loss of the volatile components of an organism, leaving a residue of carbon. It is also termed *distillation.* Fern leaves in shale in coal mines are fine examples of this delicate process.

Carboniferous. The interval of geologic time between the Devonian and Permian Periods. In North America, it is divided into the Mississippian (Lower Carboniferous) and Pennsylvanian (Upper Carboniferous) Periods. Rocks of this age contain the world's most valuable coal deposits, the Coal Measures.

Cenozoic. The fifth era of geologic time, including the present (called Recent). Its periods are the Tertiary and Quaternary. The dominant forms of life represented by fossils in Cenozoic rocks are mammals; the expansion of birds and flowering plants and the appearance of grasses are also noteworthy.

Collar. The timbering and cement work at the top of a mining shaft.

Columnar jointing. Breakage of rock into several-sided pillars. This usually occurs in basalt and is most often hexagonal. It seems to be due to shrinkage as the result of cooling. Splendid examples are to be seen in the Giant's Causeway (Ireland) and Devils Tower (Wyoming).

Complex. Containing many chemical constituents. Tourmaline and hornblende are outstanding examples of complex silicates. *Complex ore* contains several metals that are difficult to extract.

Contact

Contact. The surface where two different kinds of rocks come together. A contact is one of the best places to find minerals.

Contact metamorphism. A change in rock resulting from the introduction of molten rock into or upon older rock. Contact metamorphism may be *thermal,* involving heat only, or *hydrothermal,* involving the addition of hot solutions. Among the important minerals formed by this process are garnet (grossularite, andradite), diopside (pyroxene), epidote, zoisite, tremolite (amphibole), wollastonite, idocrase, graphite, spinel, corundum, and scheelite.

Core. The central zone of the earth's lithosphere, beneath the mantle. It consists of an outer core, which is presumably liquid, and an inner core, probably solid. It is probably composed of iron-nickel, with some silicon.

Country rock. The rock surrounding an igneous intrusion or mineral vein. The country rock may itself be mineralized to a valuable extent, or grade into the ore deposit, or be entirely barren.

Cretaceous. The last period of the Mesozoic Era, between the Jurassic and Tertiary Periods. Abundant fossils in Cretaceous rocks include ammonites, dinosaur bones, and bones of other reptiles. Rocks of this age contain extensive coal beds, and most of the world's chalk is Cretaceous. Silver, copper, lead, and zinc ores in western United States, Canada, and elsewhere are of Cretaceous age.

Crosscut. A horizontal opening in a mine, running across a vein or along the width of an ore body. It usually connects a shaft and a vein and crosses the main workings in the mine.

Crust. The solid, outer zone of the earth's lithosphere, including the continents and ocean basins, and containing the only rocks and minerals accessible to man. It extends to the Moho, or Mohorovičić discontinuity (the area of transition between the earth's crust and mantle), beneath which is the mantle.

Cupola. The upper part of a hidden body of igneous rock. The cupola of a batholith is apt to be associated with productive ore deposits of vein and other types.

Decomposition. Chemical processes of weathering. These include oxidation, carbonation, hydration, and other reactions with moisture and the gases of the atmosphere. New minerals are formed as a result, and soil is the ultimate product.

Degradation. Erosional processes. These lower the surface of the land by the erosion and transportation of rock. The opposite is *aggradation.*

Deposition. The building up of sedimentary rock by water, ice, and wind.

Desiccation. Drying up. Certain hydrous minerals desiccate when removed to an arid environment.

Devonian. The fourth period of the Paleozoic Era, between the Silurian and Carboniferous Periods. It was named from Devonshire, a county in England, where rocks of this geologic age were first studied. They contain fossils of large fishes, then the dominant life-form. Devonian iron-ore deposits are important in the southeastern corner of Canada.

Diastrophism. Movement of the solid parts of the earth. It involves *deformation* (folding) and *displacement* (faulting).

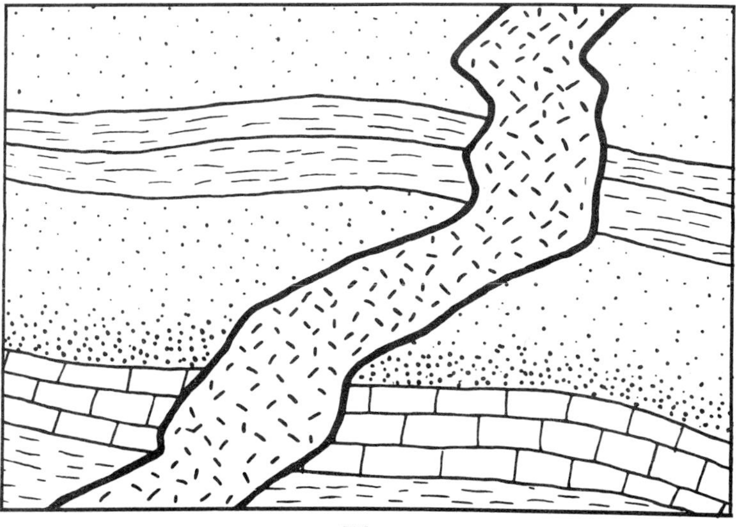

Dike

Dike. A tabular body of rock, shaped like a book or table top, cutting at any angle across older rock of any kind. This is unlike a sill, which is parallel to the adjacent rock. True dikes are of intrusive igneous origin and may be composed of a wide variety of rock types. The Yogo sapphire deposits of Montana occur in dikes. Dikes are often mineralized in association with ore veins, as at Cripple Creek, Colorado, and in Alaska and Ontario. Pegmatite dikes are the most interesting to collectors. Erosion may leave a dike standing as a rock wall. The British spelling is *dyke.*

Dip. The downward inclination of a structure or bed. The dip is measured as an angle at right angles to the strike.

Disintegration. Physical processes of weathering. These include frost action, expansion by freezing water, breakage by root growth, loosening by the movement of animals, and other ways in which rock is reduced in size without chemical change.

Drift. A horizontal opening in a mine running parallel to a vein or along the length of an ore body.

Dump. The pile of waste ore taken from a mine.

Enrichment. A secondary process (such as leaching and subsequent redeposition) involving an increase in the metal content of an ore. Copper and silver deposits are especially susceptible to enrichment.

Eocene. The second epoch of the Tertiary Period, between the Paleocene and Oligocene Epochs.

Epoch. One of the time subdivisions of a period. Epochs are of unequal length. Each epoch was characterized by its own kinds of plant and animal life, as revealed by fossils in the rocks.

Era. One of the major divisions of geologic time. The eras include Archeozoic, Proterozoic, Paleozoic, Mesozoic, and Cenozoic. They are of unequal length, and each era was characterized by its own kinds of plant and animal life, as revealed by fossils in the rocks.

Erosion. The removal of loose rock by wind, moving water, and ice. Combined with weathering and transportation, erosion (also called *degradation*) moves rock and mineral matter from one place to another. The force of gravity underlies all such activity.

Evaporite. A sedimentary rock or mineral deposited by the evaporation of water. Sea water in an enclosed basin yields this sort of product, as does concentrated lake water, such as in a playa lake. Rock salt (halite), gypsum, and anhydrite are the most common of the evaporites. These may be impure, containing admixtures of clay, sand, or carbonate matter.

Exploitation. The process of mining and marketing a mineral deposit.

Exploration. The work involved in securing information about the position, extent, and value of a mineral deposit.

Exposure. A surface outcrop of igneous rock.

Extrusion. Igneous rock that has solidified from lava, near or upon the surface. Familiar examples of extrusive rock include rhyolite, obsidian, andesite, dacite, trachyte, and basalt.

Fault. Surface along which rock has slipped. Faults may extend for miles and many be favorable places for minerals to deposit. The Comstock Lode, in Nevada, is richly mineralized for 2 miles or more along a large fault.

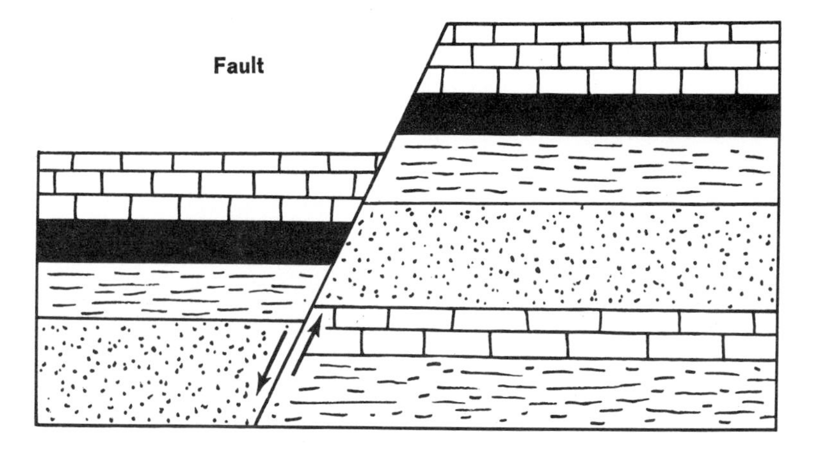

Fissure. A large fracture in rock. It may become the passageway for ore solutions that will form ore veins.

Fissure eruption. The extrusion of lava or volcanic fragments from a vent in the earth. The world's largest deposits of lava are of this type, as in Hawaii, India, the Columbia Plateau (western United States), the Paraná Basin (South America), and the North Atlantic.

Fissure vein. An ore vein occupying one or more cracks in rock. It is the most typical kind of ore deposit, yielding much of the world's metals.

Float. Loose or transported pieces of ore or rock. These have been separated from the main vein or rock by weathering or gravity. Many mineral discoveries have been made by following float to the original body.

Fold. Layered rocks that are bent into *anticlines, synclines, domes, basins,* and other structures. Most folding results from compression caused by forces acting nearly horizontally in the earth's crust. Many mineralized beds of rock have been folded. If too intensely folded, rock breaks and slips along faults.

Foot wall. The wall or rock on the underside of an inclined fault or vein.

Formation. Layers of sedimentary rock that were deposited under fairly uniform geologic conditions during a limited interval of time. They should be thick enough to be mapped and given a formation name. A formation is also a named body of igneous or metamorphic rock. Merely peculiar or interesting rocks, such as Lovers Leap and Great Stone Face, are not called formations by geologists.

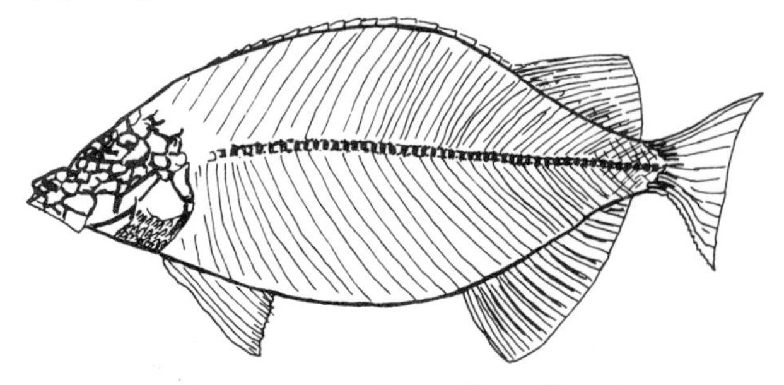

Fossil fish, Eocene; Green River, Wyoming.

Fossil. Remains or indication of ancient life. A large proportion of known fossils are mineralized by petrifaction, but actual remains of the original animal or plant may also be found, such as "refrigerated" mammoths in Siberia and Alaska and ancient logs in New Jersey.

Fracture. A crack in rock.

Fumarole. A surface opening from which volcanic gas issues. Many fumaroles deposit native sulfur and other minerals, both metallic and nonmetallic. Fumaroles are numerous in volcanic regions, such as Italy, New Zealand, and Yellowstone Park.

Gangue. The worthless material associated with ore in a vein. It is usually nonmetallic (such as quartz, calcite, fluorite, barite) but may be metallic (such as pyrite).

Gastrolith. A round, shiny stone swallowed by dinosaurs and other prehistoric animals to aid their digestion (it helped to break up swallowed food). Also called *gizzard stone.* To be sure of its identity, you should find it with fossil bones.

Geochemistry. The study of the chemistry of the earth. It includes the rocks and minerals, the waters, and the atmosphere.

Geologic time scale. The time units of geology—the *eras, periods,* and *epochs*—are of unequal lengths. The scale is always printed in reverse order as though a complete cross-section were cut through the earth; thus, the younger intervals are at the top and the older intervals are shown at the bottom, where their rocks were deposited first. Each of these terms is further explained and described in this section.

GEOLOGIC TIME SCALE

Eras	Periods	Duration of eras in years	Dominant Animal Life
CENOZOIC	Quaternary Recent Pleistocene Tertiary Pliocene Miocene Oligocene Eocene Paleocene	63 MILLION	Mammals Man
MESOZOIC	Cretaceous Jurassic Triassic	155 MILLION	Dinosaurs
PALEOZOIC	Permian Pennsylvanian Mississippian Devonian Silurian Ordovician Cambrian	375 MILLION	Reptiles Amphibians Fishes Invertebrates
PROTEROZOIC	Upper Precambrian	4000 MILLION+	Primitive multicellular forms
ARCHEOZOIC	Lower Precambrian		Unicellular forms

Geomorphology. The study of land-forms. It includes their origin and development and is thus a branch of geology.

Geosyncline. A great downfold in the earth's crust. In it sediments and lava may accumulate to many thousands of feet of thickness before being compressed and then uplifted to form folded, or *complex,* mountains. The Alps, Rockies, Urals, and other mountain systems have originated from geosynclines in this way.

Gizzard stone. A *gastrolith,* swallowed by dinosaurs and other prehistoric animals to aid their digestion.

Glory hole. A large open pit from which ore is removed through adjoining shafts.

Gossan. An iron oxide body covering one or more sulfide-bearing mineral veins or sulfide masses. It is produced by weathering.

Gradation. Erosion and deposition. Erosion and transportation are the processes of *degradation,* and deposition includes those of *aggradation.*

Ground water. The water existing within the rocks of the earth's crust, beneath the surface. It is the same as underground, subsurface, and subterranean water. It carries mineral matter in solution and deposits it to cement sediment together, to replace minerals, to petrify wood, and to form cavestone, geodes, and concretions.

Hanging wall. The wall or rock on the upper side of an inclined fault or vein.

Helictite. Twiglike cavern deposit twisted sideways, probably by currents of air.

Hydrothermal. Pertaining to hot water and its mineral deposits. These are usually of magmatic origin.

Incrustation. A coating or crust.

Intrusion. Igneous rock that has solidified from magma, beneath the surface. Familiar examples of intrusive rock include granite, pegmatite, aplite, syenite, diorite, monzonite, gabbro, and peridotite.

Jointing

Joint. One of a regular group of fractures in rock. Joints originate in igneous rock by the shrinkage of solidifying magma and lava, in sedimentary rock by the shrinkage of drying sediments, and in any kind of rock by pressure and movement resulting from the application of earth forces, or by pressure-releasing erosion. Minerals may form along the surface of a joint. Columnar jointing produces pillars in basalt and other rock.

Jurassic. The middle period of the Mesozoic Era, between the Triassic and Cretaceous Periods. It was named from the Jura Mountains, between France and Switzerland, where rocks of this geologic age were first studied. They contain fossils of the first birds and abundant fossils of ammonites (extinct cephalopod mollusks), along with dinosaurs and other reptiles. The Mother Lode gold of California is of Jurassic age, and so are iron-ore deposits of England and central Europe.

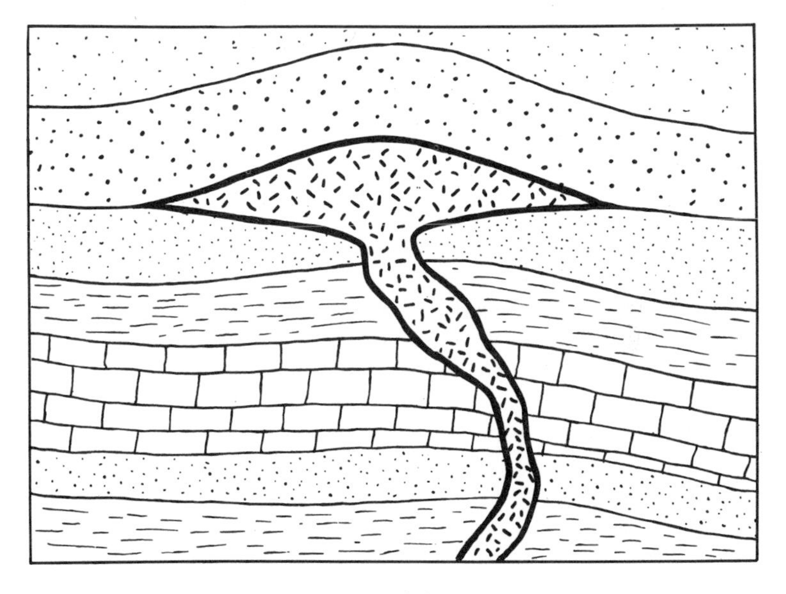

Laccolith

Laccolith. A body of igneous rock that has domed up an overlying arch of the sedimentary rock into which it has intruded. It is thus derived from a sill, which is uniformly flat. A laccolith is seldom as highly mineralized as a dike, which cuts across the adjacent rocks.

Lenticular. Lens shaped.

Level. A horizontal opening in a mine.

Lithosphere. The solid body of the earth, including the crust, mantle, and core.

Lode. A vein or close group of veins. Mining claims of lode type are staked under laws that are different from those of placer claims.

Magma. The molten matter that cools and solidifies to become igneous rock. Magma is a liquid of silicate composition, in which is dissolved steam and other gases. It contains all the natural chemical elements, and so it is the original source of all minerals formed within the earth's crust.

Mantle. The zone of the earth's lithosphere between the crust and core. At least some volcanic materials, and hence some minerals and rocks, originate in the upper levels of the mantle.

Mesozoic. The fourth era of geologic time, between the Paleozoic and Cenozoic Eras. Its periods are the Triassic, Jurassic, and Cretaceous. The dominant forms of life represented by fossils in Mesozoic rocks are dinosaurs, marine and flying reptiles, and ammonites. In this era, flowering plants, birds, and mammals made their first appearances.

Miocene. The fourth epoch of the Tertiary Period, between the Oligocene and Pliocene Epochs. Important gold and silver ores in the western United States are of this age.

Mississippian. The period of geologic time equivalent to the Lower Carboniferous. Rocks of this age contain important coal beds.

Oligocene. The third epoch of the Tertiary Period, between the Eocene and Miocene Epochs.

Ordovician. The second period of the Paleozoic Era, between the Cambrian and Silurian Periods. It was named from the Roman name for a Celtic people in Wales, where rocks of this geologic age were first studied. They contain the oldest fossils of vertebrates (fishes) and the largest fossils of trilobites. Ordovician iron-ore deposits are important in Newfoundland.

Ore. A mineral deposit containing one or more metals that can be mined or treated commercially. The useless part of ore is *gangue.*

Orogeny. Mountain making. It has nothing to do with ores except insofar as ores are often formed during the general process of mountain building. An example of an orogeny is the Caledonian orogeny, which created great mountains in Great Britain and Scandinavia during the Paleozoic Era.

Outcrop. A surface exposure of a sedimentary rock or mineral deposit.

Bedrock and overburden

Overburden. Soil or worthless rock covering a mineral deposit.

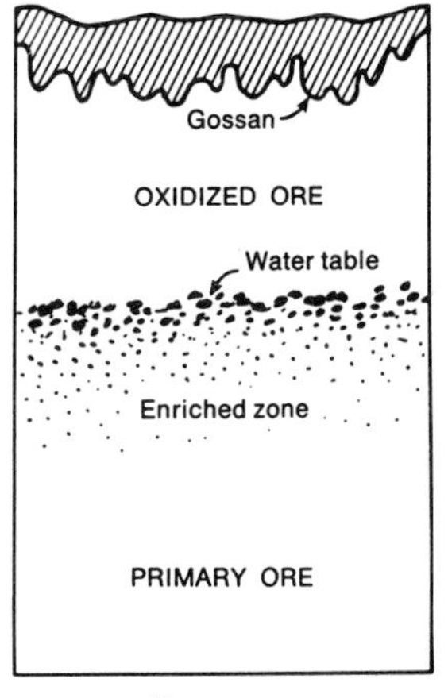

Ore zones

Oxidized zone. The portion of an ore deposit in which sulfide minerals, originally in the sulfide zone, have been changed to sulfates, oxides, and carbonates by the action of circulating air and ground water containing oxygen and carbon dioxide. This zone lies generally above the water table, in the *vadose zone,* or *zone of aeration.* The mineral specimens in this zone are typically bright colored, especially those containing copper, vanadium, and uranium.

Paleocene. The oldest epoch of the Tertiary Period.

Paleontology. The study of fossils and the ancient forms of life, whether plant or animal, which they represent. Biological evolution and geological chronology are thus aspects of paleontology.

Paleozoic. The third era of geologic time, between the Proterozoic and Mesozoic Eras. Its periods are the Cambrian, Ordovician, Silurian, Devonian, Carboniferous (Mississippian and Pennsylvanian), and Permian. The fossils in Paleozoic rocks represent the first appearance of land plants, vertebrates, amphibians, and reptiles. Invertebrates, fishes, and coal-making plants were dominant forms of life.

Pennsylvanian. The period of geologic time equivalent to the Upper Carboniferous. Its coal beds are the world's most important.

Period. One of the time subdivisions of an era. Periods are of unequal length. Each period was characterized by its own kinds of plant and animal life, as revealed by fossils in the rocks.

Permeability. The ability of a rock to transmit water under pressure. The rock must first be porous so that water can move from place to place. Mineral matter may be deposited in the pores of such rock.

Permian. The last period of the Paleozoic Era, between the Pennsylvanian and Triassic Periods. It was named from the former province of Perm, in eastern Russia, where rocks of this geologic age were first studied. The potash beds of the Permian Basin in New Mexico, Texas, and Mexico were deposited at that time. Permian salt and potash beds are important commercially in Germany and Russia, as are the Permian salt beds in Kansas and Oklahoma.

Permineralization. The filling of pores of a fossil by mineral matter dissolved in ground water. Many fossil bones and shells are petrified in this way.

Petrifaction. Conversion of organic matter to mineral matter. Most fossils are petrified, either by permineralization, replacement, or carbonation. This process is also termed *petrification.*

Placer. A loose deposit of secondary nature. It does not contain veins in place. Mining claims of placer type are staked under laws that are different from those of lode claims. Native metals, heavy minerals, and many gems occur typically in placers.

Playa. A shallow basin in a desert. Water gathers in it after a rain to make a playa lake. Playas yield deposits of potash, borax, sodium salts, and other minerals of the kind that are easily soluble and readily settle out when the water evaporates. Areas in Egypt and southwestern United States are noted for playas.

Pleistocene. The first epoch of the Quaternary Period. It is the same as the Ice Age. The most conspicuous fossils in rocks of this age are large mammals.

Pliocene. The last epoch of the Tertiary Period, between the Miocene and Pleistocene Epochs. Ore deposits of this age occur in Nevada.

Plutonic. Formed by solidification of magma deep within the earth. Plutonic bodies, or *plutons,* include those, such as batholiths and stocks, that have no visible floors. Plutonic rock is completely crystalline and typically coarse grained.

Porosity. The percentage of open space in rock or soil. A porous rock permits the deposition of mineral matter in favorable places.

Porphyry copper. A deposit of copper minerals scattered through intrusive igneous rock and surrounding rock. This is the type of large-size, low-grade ore body that furnishes most of the world's copper. The localities include such famous deposits as Bingham (Utah), Ajo (Arizona), and Chuquicamata (Chile).

Pre-Cambrian. The oldest division of geologic time. It includes the Archeozoic and Proterozoic Eras. Rocks of this age do not contain abundant fossils, and the few that are found are mostly of very primitive types of plants and animals. Pre-Cambrian rocks can be identified by their field occurrence and by radioactive dating. The vast mineral deposits of Pre-Cambrian age include hematite iron ore of Lake Superior (especially Minnesota), Ontario and Quebec, Venezuela, Brazil, and the Ukraine; magnetite iron ore of Sweden; native copper ore of Michigan; sulfide copper ore of Manitoba, Ontario, and Quebec; nickel ore of Manitoba and Ontario; tungsten ore of China; chromium ore of Rhodesia and South Africa; platinum ore of Ontario and the Soviet Union; uranium ore of Canada and the Congo; silver ore of Ontario; lead-silver ore of Australia; gold deposits of South Dakota, Ontario, and Quebec, Brazil, South Africa, India, and Australia.

Prospect. An unimproved mineral property. Further examination and testing may turn it into a working mine or else cause it to be abandoned.

Prospecting. Searching for mineral deposits.

Proterozoic. The second oldest era of geologic time, marking the last part of Pre-Cambrian history, between the Archeozoic and Paleozoic Eras. Algae and annelid worms are the main fossils in Proterozoic rocks, which must be identified by their field occurrence and by radioactive dating.

Pyritization. Petrifaction by means of iron sulfide, FeS_2. It leaves pyrite or marcasite in place of the organic matter of the fossil. This change is not stable.

Quarry. An open excavation for the removal of rock. Pegmatite deposits, building stone, and other rocks are often mined in quarries.

Quaternary. The latest period of the Cenozoic Era, including the Pleistocene Epoch and the present (called Recent). The most distinctive fossils in rocks of this age are large mammals of the Ice Age.

Raise. A vertical or inclined opening extending upward from a level.

Regional metamorphism. A change in rock resulting from crustal movements. Pressure, heat, and the introduction of fluids produce the effects that are noted in most metamorphism of this type.

Replacement. The substitution of mineral matter for organic matter in a fossil. The mineral substance is deposited from ground water. Petrified wood is the best known example of this kind of petrifaction.

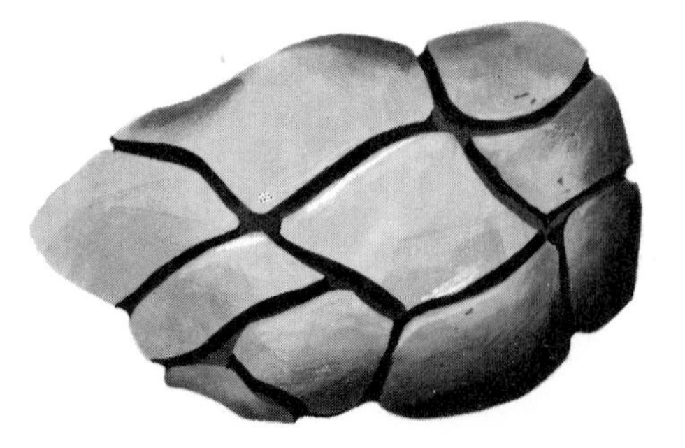

Septarian

Septarian. A concretion that has dried and cracked, the cracks being filled by veins of mineral matter. The veins may later stand up as resistant ridges or else wear down to form grooves. Collectors often mistake septaria for fossil turtles.

Shaft. A vertical or inclined opening extending downward into a mine. It may be used for entry or to remove ore or water.

Shoot. A concentration of valuable minerals in a steep, elongated vein.

Silicification. Petrifaction by means of silica, SiO_2. This is the most common process, leaving quartz, chalcedony, or opal in place of the organic matter of the fossil.

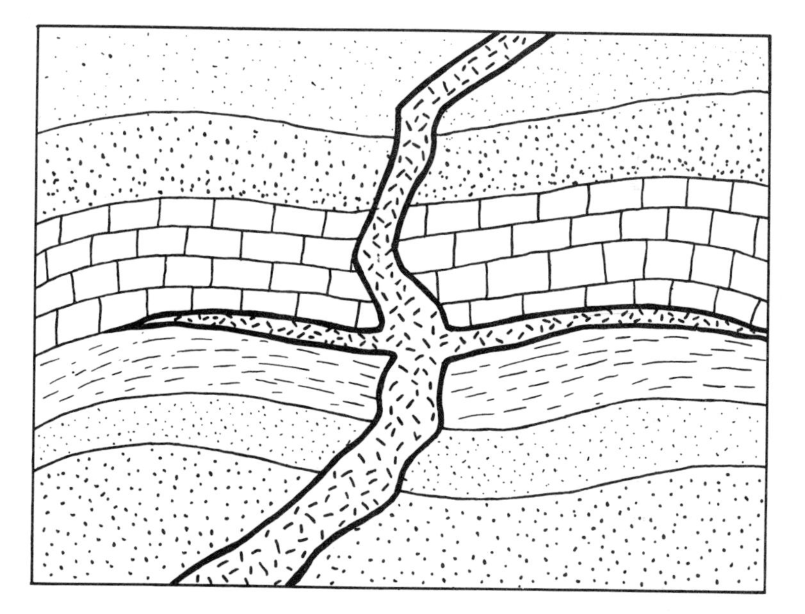

Sill and dike

Sill. A tabular body of igneous rock, shaped like a book or table top, parallel to the bedding or structure of the older rock adjacent to it. It does not have to be horizontal, although many sills are. A sill is seldom as richly mineralized as a dike, which cuts across the adjacent rocks. A sill is usually nearly uniform in thickness. When the center is domed beneath an arch of sedimentary rocks, the structure becomes a *laccolith*.

Silurian. The third period of the Paleozoic Era, between the Ordovician and Devonian Periods. It was named from the Roman name for a people in Wales, where rocks of this geologic age were first studied. They contain large fossil coral reefs and crustaceans. Silurian iron-ore deposits are important in Alabama; and Silurian salt deposits, in the lower Great Lakes region.

Soil. Broken and decayed rock and organic matter (humus). In geology (especially engineering), the loose rock and associated matter lying upon bedrock need not support plant growth in order to be called soil.

Solfatara. A fumarole yielding sulfur gases. Native sulfur is deposited around the vent.

Stalactites and stalagmites

Stalactite. Cavern deposits hanging like an icicle from the ceiling. It may join a stalagmite to form a column or pillar. Besides the familiar examples of calcite, stalactites of chalcedony, limonite, gibbsite, and other minerals are known.

Stalagmite. Cavern deposit rising from the floor. The usual mineral is calcite.

Stock. A body of intrusive igneous rock that is exposed over an area of less than 40 square miles. It is thus smaller than a batholith, which it otherwise resembles, and it may even be the upper part, or *cupola,* of a batholith not yet revealed by erosion. Many minerals occur in stocks, including veins of various metals and the porphyry copper deposits.

Stope. The underground excavation where ore is being removed in a mine.

Strike. The compass direction of a structure or bed. It is at right angles to the direction of dip.

Sulfide zone. The portion of an ore deposit in which the sulfide minerals have not been oxidized by circulating air or ground water. It lies generally beneath the water table, in the zone of saturation.

Sump. An underground excavation in a mine where water is collected for removal.

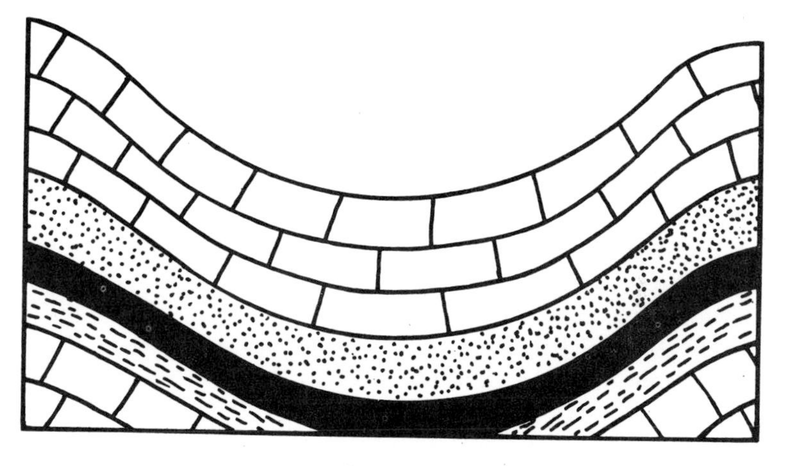

Syncline

Syncline. A fold in which the rocks have been bent downward into a trough. A *geosyncline* is a huge syncline in which sediments and lava may accumulate to a great thickness before being compressed and then uplifted to form folded, or *complex* mountains.

Tertiary. The oldest period of the Cenozoic Era, between the Cretaceous and Quaternary Periods. Its epochs are the Paleocene, Eocene, Oligocene, Miocene, and Pliocene. The dominant fossils in rocks of this age are mammals. Tertiary mineral deposits of gold, silver, copper, molybdenum, and other metals are found in mountain ranges throughout the world.

Triassic. The oldest period of the Mesozoic Era, between the Permian and Jurassic Periods. Triassic iron-ore and copper-ore deposits occur in eastern North America.

Tunnel. A horizontal passage in a mine, open at the surface at both ends. If open only at one end, it is an *adit.*

Vadose zone. The zone above the water table. It is also called the *zone of aeration,* because the pores and cracks in the rock contain both air and percolating water. In this zone is the *oxidized zone* of ore deposits and their associated minerals.

Vein. A filling of open space in rock with later minerals.

These usually come from beneath and deposit from solutions rising from magma.

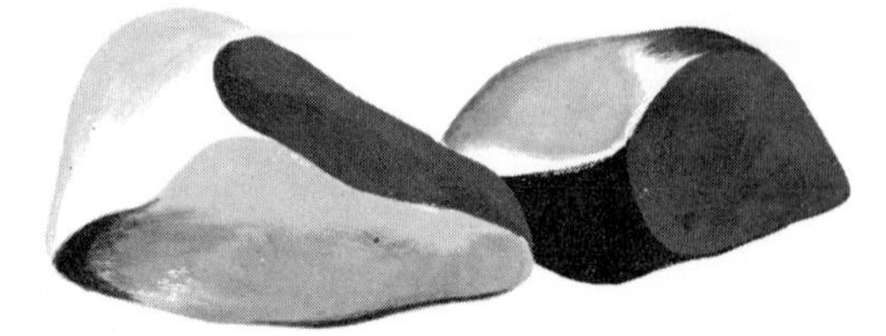

Ventifacts

Ventifact. A stone beveled by wind abrasion. It is usually glossy and pitted. A ventifact having one definite edge is an *einkanter;* having three edges, a *dreikanter.* Instead, there may be many small faces, which help to distinguish it from a gastrolith.

Volcanic neck. A cylindrical body of igneous rock occupying the vent of a former volcano. It is likely to show columnar joining.

Volcano. A central opening from which lava and other volcanic rock is erupted to build a cone around a crater. Many minerals, of which native sulfur is much the most common, form in and around volcanoes. These spectacular features of the earth's external geology occur mainly in groups and broad zones, especially the zone encircling the Pacific Ocean.

Vulcanism. Movement of molten rock and its solidification. This may take place within the lithosphere, as well as on the surface *(volcanism),* whether or not a volcano is involved.

Waste. The barren rock in a mine or that removed from it.

Water table. The boundary between the vadose zone (above) and the zone of saturation (below). Above it, the pores and cracks in the rock contain both air and percolating water. Beneath it, these spaces are completely filled with water.

Weathering. Breakdown of rock by exposure to the atmosphere and organisms. The processes are classed as *physical weathering* (disintegration) and *chemical weathering* (decomposition). New minerals form during weathering, and the ultimate result is the creation of soil, which contains rotted vegetable matter and broken and decayed rock.

Winze. A vertical or inclined opening extending downward from a level.

Xenolith. A foreign inclusion in a body of igneous rock. It is usually acquired from the country rock and may be partly dissolved (assimilated).

Zone of aeration. The zone above the water table. It is also called the *vadose zone.* The pores and cracks in the rock contain both air and percolating water. The *oxidized zone* of ore deposits and associated minerals occur in the zone of aeration.

Zone of saturation. The zone below the water table. The pores and cracks in the rock are completely saturated with water. In this zone is the *sulfide zone* of ore deposits and associated minerals.

INDEX

The index lists boldface topic headings and all illustrations in the text; also, various terms of general interest are referred to related subjects. Illustrations in color are indicated by an asterisk following the page number.